IET TELECOMMUNICATIONS SERIES 35

Series Editors: Prof. C.J. Hughes
Prof. J.D. Parsons
Dr G.L. White

T0258042

Principles of
Performance
Engineering
for Telecommunication
and Information Systems

Other volumes in this series:

Volume 9   **Phase noise signal sources** W.P. Robins
Volume 12   **Spread spectrum in communications** R. Skaug and J.F. Hjelmstad
Volume 13   **Advanced signal processing** D.J. Creasey (Editor)
Volume 19   **Telecommunications traffic, tariffs and costs** R.E. Farr
Volume 20   **An introduction to satellite communications** D.I. Dalgleish
Volume 25   **Personal and mobile radio systems** R.C.V. Macario (Editor)
Volume 26   **Common-channel signalling** R.J. Manterfield
Volume 28   **Very small aperture terminals (VSATs)** J.L. Everett (Editor)
Volume 29   **ATM: the broadband telecommunications solution** L.G. Cuthbert and
                     J.C. Sapanel
Volume 31   **Data communications and networks, 3rd edition** R.L. Brewster (Editor)
Volume 32   **Analogue optical fibre communications** B. Wilson, Z. Ghassemlooy and
                     I.Z. Darwazeh (Editors)
Volume 33   **Modern personal radio systems** R.C.V. Macario (Editor)
Volume 34   **Digital broadcasting** P. Dambacher
Volume 35   **Principles of performance engineering for telecommunication and
                     information systems** M. Ghanbari, C.J. Hughes, M.C. Sinclair and J.P. Eade
Volume 36   **Telecommunication networks, 2nd edition** J.E. Flood (Editor)
Volume 37   **Optical communication receiver design** S.B. Alexander
Volume 38   **Satellite communication systems, 3rd edition** B.G. Evans (Editor)
Volume 40   **Spread spectrum in mobile communication** O. Berg, T. Berg, J.F. Hjelmstad,
                     S. Haavik and R. Skaug
Volume 41   **World telecommunications economics** J.J. Wheatley
Volume 43   **Telecommunications signalling** R.J. Manterfield
Volume 44   **Digital signal filtering, analysis and restoration** J. Jan
Volume 45   **Radio spectrum management, 2nd edition** D.J. Withers
Volume 46   **Intelligent networks: principles and applications** J.R. Anderson
Volume 47   **Local access network technologies** P. France
Volume 48   **Telecommunications quality of service management** A.P. Oodan (Editor)
Volume 49   **Standard codecs: image compression to advanced video coding**
                     M. Ghanbari
Volume 50   **Telecommunications regulation** J. Buckley
Volume 51   **Security for mobility** C. Mitchell (Editor)
Volume 904 **Optical fibre sensing and signal processing** B. Culshaw
Volume 905 **ISDN applications in education and training** R. Mason and P.D. Bacsich

Principles of
Performance
Engineering
for Telecommunication
and Information Systems

M. Ghanbari, C.J. Hughes,
M.C. Sinclair and J.P. Eade

The Institution of Engineering and Technology

Published by The Institution of Engineering and Technology, London, United Kingdom

First edition © 1997 The Institution of Electrical Engineers
Paperback edition © 2006 The Institution of Engineering and Technology

First published 1997
Paperback edition 2006

The Institution of Engineering and Technology
Michael Faraday House
Six Hills Way, Stevenage
Herts, SG1 2AY, United Kingdom

www.theiet.org

British Library Cataloguing in Publication Data
A catalogue record for this product is available from the British Library

ISBN (10 digit) 0 86341 639 X
ISBN (13 digit) 978-0-86341-639-2

Printed in the UK by Hobbs the Printers Ltd, Hampshire
Reprinted in the UK by Lightning Source UK Ltd, Milton Keynes

Contents

1	**Introduction**		**1**
	1.1	What is performance engineering?	1
	1.2	Factors affecting performance	3
		1.2.1 User behaviour	3
		1.2.2 Natural phenomena	4
		1.2.3 Tolerances	4
	1.3	Performance criteria	4
	1.4	Applications of performance engineering	5
	1.5	Design tools	6
	1.6	References	7
2	**Analytical methods**		**8**
	2.1	Random experiments	8
	2.2	Discrete probability theory	9
		2.2.1 Fundamentals	9
		2.2.2 Conditional probability	11
		2.2.3 Statistical independance	14
	2.3	Discrete random variables	15
		2.3.1 Fundamentals	15
		2.3.2 Expectation	16
		2.3.3 Probability generating functions	17
		2.3.4 Some discrete probability distributions	19
		2.3.5 Discrete random vectors	24
		2.3.6 Independent random variables	25
		2.3.7 Compound randomness	28
	2.4	Continuous random variables	30
		2.4.1 Distribution function	30
		2.4.2 Expectation	32
		2.4.3 The Laplace transform	33
		2.4.4 Some continuous probability distributions	34
		2.4.5 Exponential distribution	35
	2.5	Random point processes	38
		2.5.1 Stochastic modelling of random events	38
		2.5.2 The Poisson process	38

2.6	Correlation and covariance	42
	2.6.1 Cross correlation	42
	2.6.2 Autocorrelation and autocovariance	43
2.7	Birth and death processes	44
	2.7.1 Definition of a birth and death process	44
	2.7.2 Behaviour of a birth and death process	45
	2.7.3 Equilibrium distribution of N(t)	47
3	**Simulation methods**	**50**
3.1	Why simulate?	50
3.2	Generating random numbers	51
	3.2.1 Midsquares method	51
	3.2.2 Linear congruential generators	52
	3.2.3 Additive congruential generators	55
	3.2.4 Computational considerations	56
	3.2.5 An example generator	57
	3.2.6 Empirical testing of generators	60
3.3	Other probability distributions	63
	3.3.1 Inverse transform	63
	3.3.2 Composition	66
	3.3.3 Accept-reject	68
3.4	Programming simulations	69
	3.4.1 Time-based simulation	70
	3.4.2 Event-based simulation	70
	3.4.3 Statistics of simulation	73
	3.4.4 Analysing and verifying simulation results	73
3.5	An example simulation programme	77
	3.5.1 The subsciber's private meter example	77
3.6	References	90
4	**Queuing systems**	**92**
4.1	Service systems	92
4.2	Performance assessment	93
4.3	Queuing models	94
	4.3.1 The arrival process	95
	4.3.2 The service facility	96
	4.3.3 Queue discipline	97
	4.3.4 Finite capacity queues	98
4.4	Kendals notation	99
4.5	Performance measures	99
	4.5.1 System performance measures	100
	4.5.2 Customer view of performance	102
	4.5.3 Calculation of performance measures	103

4.6	Little's formula	104
4.7	The Poisson arrival process	105
	4.7.1 Poisson arrivals see time averages	105
	4.7.2 Exponential service time distribution	106
4.8	The Markovian queue: M/M/r	108
	4.8.1 Steady-state distribution of N(t)	108
	4.8.2 Average delay	110
	4.8.3 Waiting time distribution	113
	4.8.4 The M/M/1 queue	114
	4.8.5 Effect of finite buffer	115
4.9	Queue with embedded Markov chain: M/GI/1	116
	4.9.1 The Pollaczek-Khinchine formula	117
	4.9.2 Waiting time distribution in M/GI/1 FCFS queue	120
4.10	Slotted M/D/1 queues	121
4.11	Some useful approximations	123
	4.11.1 Kingman's heavy traffic approximation	124
	4.11.2 Whitt's approximation	124
	4.11.3 An approximation for GI/GI/r	124
4.12	References	125
5	**Queuing networks**	**126**
5.1	Networks of queues	126
5.2	Open networks of queues	129
	5.2.1 Throughput	129
	5.2.2 Visit count	130
	5.2.3 The average total service time	131
5.3	Seperable open queuing networks	132
	5.3.1 Jackson networks	132
	5.3.2 Performance measures	133
	5.3.3 BCMP networks	134
5.4	Closed networks of queues	136
	5.4.1 Relative throughput	136
	5.4.2 Saturation point	136
5.5	Separable closed queuing networks	140
5.6	The convolution algorith	141
	5.6.1 Evaluating the normalisation constant	141
	5.6.2 Computing performance measures	142
	5.6.3 Numerical problems with the convolution algorithm	144
5.7	Mean value analysis	145
5.8	Case studies	147
	5.8.1 Central server model	147
	5.8.2 Multiprocessor systems	149
5.9	References	151

6	**Switched systems**	**152**
	6.1 Objectives of traffic engineering	152
	6.2 Traffic sources and properties	153
	6.3 Model traffic sources	154
	6.3.1 Inter-arrival times	154
	6.3.2 Number of arrivals	155
	6.3.3 Holding times	157
	6.4 Traffic process analysis	158
	6.5 Traffic problems with simple Markov model	159
	6.5.1 Infinite source and server	159
	6.5.2 Infinite source, finite server — the Erlang loss formula	161
	6.5.3 Grade of service	162
	6.5.4 Finite source and server: the Engset and binomial distributions	164
	6.6 Delay systems	166
	6.6.1 Queue with infinite buffer	166
	6.6.2 Waiting time of delayed calls	169
	6.6.3 Queue with finite buffer; blocking and delay	171
	6.7 Exchange trunking	177
	6.8 Multi-stage networks	179
	6.8.1 Non-blocking networks	179
	6.8.2 Multi-stage networks with blocking	180
	6.9 General multi-stage analysis	182
	6.9.1 Two links in tandem	183
	6.9.2 Connections other than series or parallel	185
	6.10 Multi-stage access to trunk groups	187
	6.11 Overflow traffic	189
	6.11.1 Characteristics of overflow traffic	189
	6.11.2 Dimensioning of overflow groups	192
	6.11.3 Alternative routing strategies	193
	6.11.4 Blocking with mutual overflow	195
	6.11.5 Dynamic behaviour of mutual overflow systems	198
	6.12 References	199
7	**Packet networks**	**200**
	7.1 Messages and packets	200
	7.2 Data link as a queue	201
	7.3 Poisson and near-Poisson queues	203
	7.4 Pooled point processes	203
	7.5 Kleinrock's network delay bound and optimisation	208
	7.6 Network delay and routing	213
	7.7 Finite buffer capacity	214
	7.7.1 Uniform loss network (hypothetical)	214
	7.8 Packet length	217

7.9	Data link repeat protocol	218
	7.9.1 Stop and wait protocol	219
	7.9.2 Go-back-N	220
	7.9.3 Select-repeat protocol	223
7.10	Flow control	225
	7.10.1 Simple model for flow control analysis	226
	7.10.2 Parallel virtual circuits with flow control	230
	7.10.3 Flow control: further factors	233
7.11	ATM	235
7.12	Modelling traffic sources	240
	7.12.1 Data sources	240
	7.12.2 Speech sources	240
	7.12.3 Video sources	242
7.13	Case studies	246
	7.13.1 Case study 1: a token ring	246
	7.13.2 Case study 2: speech on ATM links	250
7.14	References	256
8	**Introduction to reliability**	**259**
8.1	Reliability performance	259
	8.1.1 Cost effectiveness	259
8.2	Reliability and availability	262
8.3	Limitations of reliability estimation	264
8.4	Basic relationships	264
	8.4.1 Reliability functions	264
	8.4.2 Failure rate	265
	8.4.3 Mean time to failure	267
	8.4.4 Failure conditions	268
8.5	Reliability modelling	269
	8.5.1 Constant failure rate	270
	8.5.2 Lifetime PDF having a normal distribution	270
	8.5.3 Weibull distribution	271
8.6	Components	273
8.7	Estimating component reliability	274
	8.7.1 Exponential distribution (constant failure rate)	274
	8.7.2 Weibull distribution	276
8.8	Accelerated life testing	279
8.9	Confidence intervals	280
8.10	Reliability standards	281
8.11	Burn-in	281
8.12	System reliability assessment	282
	8.12.1 Multi-component systems	283
	8.12.2 Maintenance spares	284
	8.12.3 Relaibility block diagrams	285

		8.12.4 Majority decision redundancy	287
	8.13	Markov modelling	288
		8.13.1 Parallel redundancy (hot standby)	288
		8.13.2 Cold standby operation	290
	8.14	Fault tree analysis	291
	8.15	Repairable systems	291
		8.15.1 Availability	294
	8.16	Redundant systems with repair	294
		8.16.1 Parallel redundancy with repair	294
		8.16.2 Majority decision redundancy	295
	8.17	Software reliability	298
		8.17.1 Software testing	299
	8.18	References	301
9		**Miscellaneous examples of performance engineering**	**302**
	9.1	Introduction	302
	9.2	Tolerances	302
		9.2.1 Component values	303
		9.2.2 Timing tolerances	305
	9.3	Transmission systems	307
		9.3.1 Repeated transmissions	308
		9.3.2 Performance of error detecting and correcting codes	312
	9.4	Mobile radio example	314
	9.5	References	319

Foreword

The IEE's Telecommunications Series is aimed chiefly at those engineers who are near the beginning of their careers or who are moving to a new specialised field. In keeping with this approach, it is expected that the reader of this work will have some knowledge of the basic principles of telecommunications but will not have had extensive experience of teletraffic theory.

In the early days, the field of teletraffic engineering was covered by specialised mathematical works by authors such as Syski and Benes. These provided excellent support for the expert but were rather heavy going for the newcomer. An early contribution to the Telecommunications Series, 'Principles of telecommunication traffic engineering' by the late Donald Bear, was first published in 1976 and gives an excellent introduction to the subject. However, there have been some important developments in the last 20 years.

First, switched telecommunication networks have changed radically. Step-by-step systems have virtually disappeared in most countries, so a study of trunking and grading systems has little practical importance. Matrix systems in both the space and time domains have developed and there is a move away from circuit switching to a range of packet systems, not only for data but also for the multiservice networks of the future.

The second major change has been in the applications of queuing theory. Kendall's work on this subject dates from the early 1950s and has provided valuable tools for the study of teletraffic performance. The theories of embedded Markov chains and of birth and death processes have been developed extensively and have enabled the behaviour of many queuing processes to be understood more clearly.

Finally, there has been an increasing realisation that teletraffic engineering has much to offer and to receive from other specialised fields. This has resulted in the development of a newly defined area of study which has now become known as 'performance engineering'.

In this book, after a brief introduction, we present a fundamental review of some basic probability and statistical theory in Chapter 2. Most readers will have some knowledge of probability theory and so parts of this chapter may be treated as a revision exercise. However, some specific aspects are developed in somewhat greater depth to serve as a foundation for the understanding of more advanced work. Computer simulation (Chapter 3) has now become of major importance in performance engineering, whereas a few years ago it was regarded as a very expensive tool to be used only by those with access to large mainframe computers. The power available with even a modest personal computer is increasing rapidly and the use of a computer program to replace analysis becomes almost too tempting. The case of the student who wrote a program based on Simpson's rule to evaluate the integral of a simple exponential function is an extreme example but, in our experience, there have been many cases of simulation over-use. On the other hand, some systems are now becoming so complex that simulation becomes the only feasible approach to performance estimation. However, it is important that some attempt at analysis, even on a greatly simplified system, should be made to support the simulation wherever possible.

In Chapters 4 and 5 the relevant theories of queues and networks of queues are discussed. It is here that we see the first extension of performance engineering beyond the strict confines of teletraffic. In a very wide range of engineering systems, scheduling queues are formed and the behaviour of such queues is a fundamental factor in determining the performance of the system.

Chapters 6 and 7 provide the application of the theory to circuit- and packet-switched systems. These chapters may seem somewhat brief in relation to the rest of the book but, in a work of this size, it would be impossible to treat every possible telecommunications application. It is more important to understand the underlying theory and allow the reader to apply it to specific systems. Case studies and examples are included to indicate the general approach.

The subject of reliability is now of such importance in all telecommunication systems that we have given it a separate chapter (Chapter 8). Although the terminology is different, the remarkable similarity to teletraffic theory will be apparent.

Finally, Chapter 9 provides a few examples of the application of performance engineering to other related fields. A full treatment of the many applications of performance engineering is not possible in a book of this size but the examples illustrate its potential. Of particular interest is subject of cellular mobile systems, where subjects as diverse as radio fading and teletraffic interact intimately to determine the performance of the system.

In a book of this nature we must acknowledge our debt to several colleagues at the University of Essex and to delegates at IEE and International Teletraffic Conferences for helpful discussions. Thanks are also due to late Mr D. K. N. Hughes for some useful suggestions and checking the maths and to Mrs P. A. Crawford for valuable secretarial help.

Chapter 1

Introduction

1.1 What is performance engineering?

Any engineered system should be designed to have a 'performance'. The performance requirements should be stated at the outset and the system should be designed to meet the performance specification. This would seem to be a relatively straightforward concept (although it may be difficult in practice to achieve the desired performance). Suppose, for example, it is required to design a car that will accelerate from 0 to 60 mph in 7.3 seconds. What exactly do we mean? Do we mean, for example:

(a)   all production cars when driven by competent drivers will reach 60 mph in 7.3 seconds or less

or (b)   a test sample of 10 cars off the production line driven by works drivers all reached 60 mph in 7.3 seconds or less

or (c)   a test sample of 10 cars off the production line driven by works drivers all reached 60 mph in an average time of 7.3 seconds

or (d)   calculations based on torque-speed curves, air resistance etc. show that, if all gear changes are made at exactly the right engine speeds, the car should reach 60 mph in 7.3 seconds.

Of course, for a system which is manufactured on a highly automated production line, such as a motor car, the variations in performance between individual units may be expected to be relatively small. However, manufacturing tolerances will inevitably produce some differences in performance from unit to unit. Most car buyers would assume that (a) was the intention and would be satisfied with the performance of their particular car. If the performance were determined under condition (b) this would appear satisfactory but, with such a small sample, there is always the possibility of a 'rogue' unit that falls below the required standard. In case (c), it might be assumed that about half the cars produced would have a performance worse than expected. Case (d) might raise questions as to the truth of the claim, since making the gear changes at exactly the right engine speeds would be beyond the capabilities of even professional drivers. A motor car is an example with relatively small performance variations but other systems can have a much wider range of performance. The classic case is a telephone exchange, in which the performance as perceived by a particular user is greatly affected by the behaviour of other users. The first paper published by A. K. Erlang in 1909 may be regarded as the beginning of performance engineering and, although the range of the subject has been greatly extended since then, it contains the essential features of performance engineering, namely:

	(i)	the estimation of performance is based partly on statistical data and/or assumptions
and	(ii)	some aspects of the estimated performance are stated in terms of probabilities.

The performance engineering design process is illustrated in Figure 1.1.

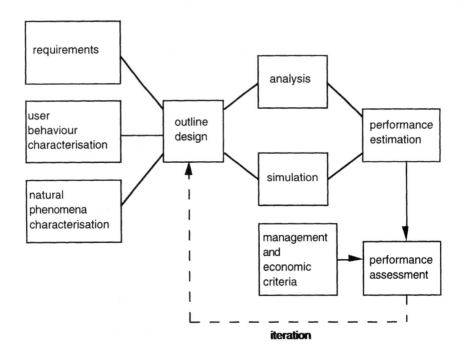

Figure 1.1   Performance engineering design process

1.2   Factors affecting performance

1.2.1   User behaviour

Many factors can affect the performance of a large system. Since most systems are used by humans, it is to be expected that human behaviour will play an important part. Even single-user systems depend to some extent for their performance on the human factor. For example, in the motor car example given above, the acceleration actually achieved will depend on how competently the driver makes the gear changes. Human behaviour is even more important in multi-user systems, which form the foundation of performance engineering. A considerable body of expertise has been built on the assumption that calls will be made at random from the relatively large number of subscribers connected to a telephone exchange. This is usually justified in practice but, in order to carry out performance calculations, it is necessary to justify a value for the average number of calls made over a period of time (the so-called 'busy hour'). For a new system, measurements can often be made on existing systems to provide data on user behaviour under broadly similar conditions. Even then, it is not usually practicable to plan for unforeseen circumstances. For example, a sudden deterioration in weather conditions or a sudden fluctuation of financial markets might result in greatly increased traffic for a short time. Furthermore, performance engineering is usually carried out on systems that are about to be built. Under growth conditions, common in telecommunication systems, an estimate has to be made of the traffic to be experienced some years into the future. This in turn is dependent on a wide range of social and economic circumstances

Beyond the narrow field of telephone calls made over narrow bandwidth channels, more advanced telecommunication systems are affected in different ways by the characteristics of the users and need to be dealt with in different ways. The characteristics of packet-switched data channels are relatively straightforward, although even here problems arise when there are intermittent transmissions of massive amounts of data, as in computer-aided design systems. The trend towards some form of packet switching for services such as speech and video has led to a need to specify new aspects of the user input for such channels. The characteristics of both continuous and conversational speech are well understood, although it may be noted that most of the experimental work has been based on European languages, particularly English. When we consider video systems, the situation becomes very complex. One way of digitally encoding video signals is to transmit the differences between one picture frame and the next. The rate of data transmission will then depend on the amount of movement in the scene and this is even more difficult to characterise in statistical terms. Although a 'typical' telephone conversation can be envisaged, in the sense that most telephone conversations will be very similar, it is not easy to imagine a typical video sequence.

1.2.2 Natural phenomena

In telecommunication and information systems, some natural phenomena such as noise and radio fading can only be described in statistical terms. Considerable knowledge exists about the behaviour of digital transmission systems in noisy and fading environments, and performance has usually been expressed in terms of bit error ratios (BERs). However, if error detection/correction is applied to the system, it may be necessary to estimate the burst characteristics of the errors as well as the BER.

In some systems, there is a close relationship between the user's input to the network and the natural phenomena. In a cellular mobile radio system, for example, the statistical analysis of the handovers between cells will depend both on the fading phenomena and the distribution of vehicle velocities in the cells.

1.2.3 Tolerances

Manufacturing tolerances play a part that may be of critical importance for the performance of some systems but may be virtually ignored in others. Generally, electronic systems are very tolerant of manufacturing tolerances, although the time for a component to operate may affect performance if it is much greater than the average. In some large digital systems, for instance, the spread of timing delays in the switching gates imposes severe limits on the system design.

1.3   Performance criteria

Even if it is possible to characterise the factors affecting the performance of a system, the design cannot be carried out until the required performance of the complete system has been specified, at least in general terms. This is not always easy since the users of the system may not be able to define, in statistical terms what they require. Car manufacturers decide what performance they would like to achieve for a particular model and the success of their surveys/guesswork is determined by the sales achieved. Economic criteria are also important for multi-user systems. For example, the safety system of an airline might be designed on the basis that just sufficient passengers may be killed each year so as not to produce a significant fall in ticket receipts! One hopes that most airlines adopt a more responsible attitude to human safety than this but, nevertheless, all airlines make an explicit or implicit decision on how much they are prepared to spend on those aspects of the system that affect passenger safety.

Operators of telecommunication systems have devoted considerable effort in attempts to define performance criteria. One economic approach is, for example, to provide additional trunks to carry traffic during peak periods until the point is reached at which the cost of provision of an additional trunk just balances the revenue derived from that trunk (Moe's principle) [1]. This gives the best possible service to the users of the system without the telephone administration incurring a cost penalty. However, in a competitive environment, this may not lead to the highest profitability. A different approach would be to plot the proportion of telephone calls that cannot be connected against the complaints received. The performance limit is then set at the point at which there is a significant increase in the complaint level. Of course, judgement has to be exercised in deciding the complaints level to be selected. It has been found that even a 'perfect' performance would be unlikely to produce zero complaints!

For multi-user systems 'human factors' experiments may be carried out to assess the effects of impairments in a communication channel. A group of (usually) naive observers are asked to judge the performance of the channel when impairments are introduced under controlled conditions. The process is not as simple as it might appear. Some people are more critical under artificial test conditions than they would be as day to day users of the system. The duration of a test run is important since an impairment introduced in a presentation lasting a few seconds may be more noticeable than when it is introduced during a test lasting several minutes. Nevertheless, human perception tests are invaluable in providing an input to management decisions on the performance criteria for the system. The other vital input to the management process is the economic aspect. For example, some telecommunications operators in the USA have found that a reduced bandwidth telephone channel, although producing a significant loss of quality, is still acceptable in view of the reduced cost.

1.4 Applications of performance engineering

The basic philosophy of performance engineering has many applications as may be seen from some of the examples given above. In telecommunication and information systems, the early work carried out by Erlang [2] and others for switched telephone networks has been greatly extended. Packet-switched systems, for instance, are becoming important for real-time services such as voice and video in addition to data, and the performance of such services has to be judged in terms of factors such as packet delay.

In multi-user computer systems, an important aspect of the performance is the queueing delay for common data servers. The design of cellular mobile radio systems is possibly one of the most complex examples of performance engineering. The multi-user input depends not only on the total traffic generated but also on the positioning of the mobile units, and relevant natural phenomena include the extent and rates of fading on the radio channels. This affects the performance of both the communication and the signalling channels used to control the system.

A more general application of performance engineering techniques is in designing the reliability of a system. The failure mechanisms of most electronic components are now well understood and so the rates and modes of failure of complex systems can be determined with comparative accuracy. More importantly, the weak points in a system can be identified and, if necessary, the reliability can be improved to the required level.

1.5   Design tools

Many of the design tools for performance engineering have been developed over a long period. The early work by Erlang [2], for example, dates from the beginning of the century. Queuing theory has been developed extensively and a considerable amount of information is now available to designers. More general work on mathematical statistics and probability theory is available in textbooks. In this book we will draw extensively on such theory and indicate how it can be applied to telecommunication and information systems. However, there often remains the problem of characterising the user inputs in sufficiently simple terms to enable the analysis to proceed. As we have seen, this is not always a simple process. Then, assuming the user inputs have been dealt with, the main problem facing the system designer is not so much the lack of relevant mathematical theory but how to apply the theory to the system being designed.

Although analysis has proved a very useful tool in the past, many of the systems coming into operation today are so complex that analysis alone is not sufficient to estimate the performance even approximately. The availability of relatively inexpensive computer power has provided the performance engineer with a means of estimating the performance of complex systems without the expense of constructing the system only to find that it does not perform as well as expected. Computer simulation overcomes many of the problems encountered in system analysis. The problem of characterising the user inputs remains as before although, in simulations, it may not be necessary to use simplified distributions that facilitate mathematical manipulation. It is quite possible to use the actual measured distributions.

There are some important disadvantages in the use of computer simulation. First, it is important to ensure that the simulation is truly representative of the system under consideration. It is possible to ignore what appear to be minor aspects of a system that in fact have a substantial effect on the performance. The other difficulty in using simulation techniques is in estimating the accuracy the results. It is possible to carry out a perfectly valid simulation run but, because of the random statistics, the results are not representative of the true performance of the system. Statistical techniques are available for estimating the probable accuracy of the results but, even then, there is a non-zero probability that the results are unrepresentative.

Overall, the most satisfactory approach is to combine both analytical and simulation techniques. This may involve simplifying the system down to a point where analysis is possible. If the simulation results then agree with the analysis for such a simplified system, there is a very good chance that the simulation of the full system will enable the performance of the completed system to be predicted with sufficient accuracy.

1.6 References

1 LITTLECHILD, S.C.: 'Elements of telecommunications economics' (Peter Peregrinus, 1979) pp. 151–157

2 ERLANG, A. K.: 'Solution of some problems in the theory of probabilities of significance in automatic telephone exchanges' *Post Office Electrical Engineers Journal,* 1918, **8**, pp. 33–45

Chapter 2

Analytical methods

2.1 Random experiments

Most people are familiar with the basic concepts of probability. For example, if a die is tossed 600 times, we would expect a six to occur about 100 times. Intuitively, we know that the relative frequency or probability is 1/6. Of course in the experiment, we would not always expect to get exactly 100 sixes. The sample probability would not be exactly 1/6 but in this case we know the correct answer if the die is unbiased.

In another example suppose that after transmission over a particularly noisy binary digital communication channel, an average proportion of 3 bits per 100 are interpreted incorrectly at the receiver. We say that the transmission errors occur with a probability of 0.03. Let us now confine our attention to one particular digit in a sequence of bits about to be transmitted over the channel. When this particular bit reaches the receiver, we distinguish only two possible outcomes of this experiment:

(a) the bit is received with the wrong value

(b) the bit is received with the correct value

Intuitively, we suspect that outcome *b* is more likely to occur than outcome *a*, but until the digit is received the actual outcome of the experiment is uncertain. In the absence of a detailed knowledge of all the influencing factors, we hypothesise that the outcome of the experiment is governed by chance. Thus, we model the transmission of the digit over the channel by a random experiment with just two possible outcomes *a* and *b* ; we assign a probability of 0.03 to outcome *a* and a probability of 0.97 to outcome *b*. The outcome for any particular digit is unpredictable but in a long series of digits, we should expect the proportion of bit errors to be close to 3%.

Random experiments are idealised models of real systems. The possible outcomes of the experiment are assigned probabilities which as far as possible reflect their relative frequencies in the real system. Probability theory provides a formal mathematical framework for manipulating the relative frequencies in the guise of statistics to obtain further information about the system.

2.2   Discrete probability theory

2.2.1   *Fundamentals*

The theory of probability is based on the concept of an arbitrary repeatable random experiment. This abstraction enables several important fundamental results to be developed. These results may then be used in any application once an appropriate random experiment has been identified.

The sample space, $\Omega = \{\omega_0, \omega_1, \omega_2, \dots \}$ is the set of all possible elementary outcomes of the random experiment. One, and only one, elementary outcome or sample point, ω_i may occur at each repeat of the experiment. In the case of the die, $\Omega = \{1,2,3,4,5,6\}$.   The same random experiment may give rise to several different sample spaces according to the problem at hand. For example, when a bit is transmitted over a channel, an analysis of the balance of the coding algorithm would consider the sample space

$$\Omega = \{\text{'0 transmitted', '1 transmitted'}\}$$

whereas an error analysis would consider the sample space

$$\Omega = \{\text{'transmission error', 'no error'}\}.$$

Subsets of Ω are called events. Thus an event $A \subseteq \Omega$ ('A is a subset of Ω') is a collection of sample points. $\varnothing$ (the empty set) is the 'impossible' event and Ω is the certain event. If A is an event (i.e. $A \subseteq \Omega$), then $\overline{A} = \{\omega \varepsilon \Omega \mid \omega \varepsilon A\}$ is the event complementary to A. Clearly, in any repeat of the random experiment, either $\overline{A}$ or A must occur and $\overline{A}$ can occur only if A does not occur. In other words, $\overline{A}$ and A are mutually exclusive and exhaustive events. For example in the die throwing experiment, let

$$A \quad = \quad \{2,4,6\} \quad \text{(the number thrown is even)}$$

$$B \quad = \quad \{1,2\} \quad \text{(the number thrown is less than 3)}$$

Then $$\overline{A} \quad = \quad \{1,3,5\}$$

$$A \cap B \quad = \quad \{2\}$$

and $$A \cup B \quad = \quad \{1,2,4,6\}$$

The probability measure is a function $P:\Omega \rightarrow [0,1]$ which assigns a real numerical value to each event (that is, to each subset of Ω). The value assigned to the event $A \subseteq \Omega$ is written $P(A)$ and is referred to as the probability of event A. The values assigned by P are to some extent arbitrary but they must be consistent with the following axioms:

(i) $P(A) \geq 0$ for every event A

(ii) $P(\Omega)$ = 1

(iii) If A and B are mutually exclusive events (i.e. $A\&B = \emptyset$)   then

$$P(A\cup B) = P(A) + P(B)$$

(iv) If A_1, A_2, A_3, ... are mutually exclusive events then

$$P(A_1 \cup A_2 \cup A_3 \ldots) = P(A_1) + P(A_2) + P(A_3) + \ldots$$

$$= \sum_{i=0}^{\infty} P(A_i)$$

In discrete probability a (possibly zero) probability $P(\omega_i) \geq 0$ is assigned to each sample point ω_i . Since $\Omega = \{\omega_0\}\cup\{\omega_1\}\cup\{\omega_2\}\cup \ldots$ and the sample points are mutually exclusive, it follows from axioms (ii) and (iv) that:

$$\sum_{i=0}^{\infty} P(\omega_i) = P(\Omega) = 1$$

If A is an event, then similarly

$$P(A) = \sum_{\omega_i \varepsilon A} P(\omega_i)$$

(2.1)

It follows from the axioms that:

$$P(\emptyset) \ = \ 0$$

$$P(\overline{A}) \ = \ 1 - P(A)$$

$$P(A \cup B) \ = \ P(A) + P(B) - P(A\&B)$$

$$A \subset B \ \Rightarrow \ P(A) \leq P(B)$$

In the case of the die, each sample point in $\Omega = \{1,2,3,4,5,6\}$ should be assigned the probability measure 1/6. Thus if $A = \{2,4,6\}$ and $B = \{1,2\}$, then

$$P(A) \ = \ P(2) + P(4) + P(6) \ = \ 1/6 + 1/6 + 1/6 \ = \ 1/2$$

$$P(B) \ = \ P(1) + P(2) \ = \ 1/6 + 1/6 \ = \ 1/3$$

$$P(\overline{A}) \ = \ P(1) + P(3) + P(5) \ = \ 1/2 \ = \ 1 - P(A)$$

$$P(A\&B) \ = \ P(2) \ = \ 1/6$$

$$P(A \cup B) \ = \ P(1) + P(2) + P(4) + P(6) \ = \ 2/3$$

2.2.2 *Conditional probability*

It is often useful to calculate the probability that the event A occurred on a particular repeat of the random experiment in which it is known that some other event B also occurred. In other words, if all that is known about the outcome ω is that $\omega \varepsilon B \subseteq \Omega$, what is the probability that $\omega \varepsilon A$? This probability is known as the conditional probability of the event A given that event B occurred and is written $P(A|B)$. For example, in the die experiment where $A = \{2,4,6\}$ and $B = \{1,2\}$, $A\&B = \{2\}$. Hence, if it is known that B occurred, that is $\omega = 1$ or 2, then A would also have occurred on the same trial only if ω was in fact 2. Since 1 and 2 are equiprobable outcomes, the conditional probability of a given B must be

$$P(A|B) \ = \ 1/2$$

Essentially the knowledge that B did occur allows us to reduce the sample space to the subset B when examining other possible consequences of this particular trial. Now if $\omega \varepsilon B$ then $\omega \varepsilon A$ if, and only if, $\omega \varepsilon A\&B$. This leads us to define the conditional probability of A given B to be

$$P(A|B) \; = \; \frac{P(A\&B)}{P(B)}$$

(2.2)

It follows that

$$P(A\&B) \; = \; P(A|B) \, P(B)$$

$$P(A|B) \; = \; \frac{P(B|A) \, P(A)}{P(B)}$$

(2.3)

The latter proposition is known as Bayes' theorem.

Conditional probability provides a useful means of calculating a probability whose value is not immediately apparent by splitting the event concerned into subsidiary events. Since $A = (A\&B) \cup (A\&\bar{B})$, and the union is disjoint (i.e $(A\&B)\&(A\&\bar{B}) = \varnothing$), it follows from axiom (iii) that

$$P(A) \quad = \; P(A\&B) \, + \, P(A\&\bar{B})$$

$$= \; P(A|B)P(B) \, + \, P(A|\bar{B})P(\bar{B})$$

> *Example: A binary communications channel employs a coding technique which ensures that the bit values 0 and 1 are transmitted in the relative proportions 4:3. Owing to noise, one in four of all 0's transmitted are received as 1, and one in five of all 1's transmitted are received as 0. What is the probability that a bit received as 0 was originally transmitted as 0?*

In probabilistic terms, the problem may be stated as follows:

$$P(0 \text{ sent}) = 4/7 \qquad P(1 \text{ sent}) = 3/7$$

Owing to noise,

$$P(1 \text{ received} | 0 \text{ sent}) = 1/4$$
$$P(0 \text{ received} | 1 \text{ sent}) = 1/5$$

We seek

$$P(0 \text{ sent} \mid 0 \text{ received}) \quad = \quad \frac{P(0 \text{ sent } \& \text{ 0 received})}{P(0 \text{ received})}$$

To compute the denominator we first split it

$$P(0 \text{ received}) \quad = \quad P(0 \text{ sent } \& \text{ 0 received}) + P(1 \text{ sent } \& \text{ 0 received})$$

Now $P(0 \text{ sent } \& \text{ 0 received}) = P(0 \text{ received} \mid 0 \text{ sent}) \, P(0 \text{ sent})$

$$= \quad (1 - P(1 \text{ received} \mid 0 \text{ sent})) \times P(0 \text{ sent})$$

$$= \quad (1 - 1/4) \times 4/7 \quad = \quad 3/7,$$

and $P(1 \text{ sent } \& \text{ 0 received}) = P(0 \text{ received} \mid 1 \text{ sent}) \, P(1 \text{ sent})$

$$= \quad (1/5) \times (3/7) \quad = \quad 3/35.$$

Thus $P(0 \text{ received}) \quad = \quad 3/7 + 3/35 = 18/35,$

(observe that $P(0 \text{ received}) \ne P(0 \text{ transmitted})$)

hence, $P(0 \text{ sent} \mid 0 \text{ received}) \quad = \quad (3/7) \div (18/35) = 5/6.$

The principle of splitting up events is generalised in the theorem of total probability.

Let $B_1, B_2 \ldots , B_n$ be events such that

(a) $B_i \, \& \, B_j = \varnothing$ if $i \ne j$, i.e. no two events can both happen at the same trial; they are mutually exclusive;

(b) $B_1 \cup B_2 \cup \ldots \cup B_n = \Omega$, i.e. one or other of the events must happen at each trial, they are mutually exhaustive;

(c) $P(B_i) > 0$ for $i = 1,2, \ldots ,n$, i.e. each event is 'possible'.

Then, for any event A

$$P(A) = \prod_{i=1}^{32} P(A_i) = (0.97)^{32} = 0.377$$

Therefore, the probability that a block contains one or more errors is

$$P(\overline{A}) = 1 - P(A) = 0.623$$

Thus, over 60% of blocks contain one or more errors! If incorrectly received blocks must be retransmitted complete. Furthermore, a retransmitted block is still liable to error. This calculation shows the necessity to maintain a very low bit error ratio in a digital communications system. In practice, errors due to noise would tend to occur in bursts and therefore the independence assumption might not bear close scrutiny. However, without it the calculation of block error ratio is much more difficult.

2.3 Discrete random variables

2.3.1 Fundamentals

In simple applications such as the die throwing experiment, the sample points are readily identified and assigned a probability measure. In more complex situations, however, the sample points may partition the sample space more finely than necessary for defining the events of practical interest. In other words, each event may consist of many sample points whose individual probabilities are of little importance and would be tedious to list. Random variables are often used to distinguish the events of interest more succinctly.

A random variable is a function which associates a real number with each outcome of a random experiment, i.e. it is a mapping $X : \Omega \rightarrow R$. For example, consider two exchanges in a telephony network connected by a full-availability group of n circuits. The occupancy of the group may be described by the n-vector $(S_1, S_2, \ldots, S_n)$, where S_i takes the value 0 or 1 depending on whether the ith circuit is free, or busy carrying a telephone conversation. In the random experiment consisting of examining the state of the group at some time, the sample space of all possible n-vectors would contain 2n sample points. For dimensioning the route, however, we need only be concerned with the total number of busy circuits, X say, which would reduce the size of the state space to n+1. In this example, the random variable $X : (S_1, S_2, \ldots, S_n) \rightarrow S_1 + S_2 + \ldots + S_n$.

A discrete random variable X has a finite or countably infinite image set of distinct values which it may assume: $X(\Omega) = \{x_0, x_1, x_2, \ldots\}$. The most elementary event concerning X takes the form

$$'X = x_i' \quad = \quad \{\omega \varepsilon \Omega \mid X(\omega) \quad = \quad x_i\}$$

The probability of this event is denoted by $p_i = P(X = x_i)$ for $i = 0, 1, 2, \ldots$ The sequence $\{p_0, p_1, p_2, \ldots\}$ is called the discrete probability distribution of the random variable X. It follows from the axioms satisfied by the underlying probability measure P that $p_i \geq 0$ and

$$\sum_{i=0}^{\infty} p_i \quad = \quad P\{X(\omega) \varepsilon X(\Omega)\} \quad = \quad P(\omega \varepsilon \Omega) \quad = \quad 1$$

$$(2.6)$$

Any sequence of non-negative real numbers which sum to unity defines a discrete probability distribution. Some common probability distributions will be dealt with later in this chapter.

2.3.2 *Expectation*

If X is a random variable, and $\phi: R \to R$ is any function, then $\phi(X)$ is a mapping from Ω to R, i.e. is another random variable. If X is a discrete random variable, then so too is $\phi(X)$, and we define the expectation of $\phi(X)$ as

$$E(\varphi(X)) \quad = \quad \sum_{i=0}^{\infty} \varphi(x_i) p_i$$

$$(2.7)$$

In particular, the expectation of X itself is

$$E(X) \quad = \quad \sum_{i=0}^{\infty} x_i p_i$$

In a long series of trials of the underlying random experiment, we would expect $X(\omega)$ to assume the value x_i with a relative frequency close to p_i. Thus, $E(X)$ represents the limiting value of the average of an infinite series of such samples of X. $E(X)$ is called the mean of X.

Example: in the fair die experiment, find the expectation of X.

$P(X = i) = 1/6$ for $i = 1,2,3,4,5,6.$

$$E(X) = 1.\frac{1}{6} + 2.\frac{1}{6} + 3.\frac{1}{6} + 4.\frac{1}{6} + 5.\frac{1}{6} + 6.\frac{1}{6} = 3.5$$

The following are some properties of the expectation operator:

(i) $X \geq 0$ $\Rightarrow$ $E(X) \geq 0$

(ii) $X = c$ (a constant) $\Rightarrow$ $E(X) = E(c) = c$

(iii) c constant $\Rightarrow$ $E(cX) = c\,E(X)$

(iv) $E[\phi_1(X) + \phi_2(X)]$ $=$ $E[\phi_1(X)] + E[\phi_2(X)]$

The expectations of higher powers of X are known as moments, $E(X^n)$ being the nth moment of X. The variance of X is a modified second order moment which indicates to what extent the distribution of X is dispersed around its mean value.

$$
\begin{aligned}
Var(X) &= E(\{X - E(X)\}^2) \\
&= E(X^2 - 2\,E(X) + \{E(X)\}^2) \\
&= E(X^2) - \{E(X)\}^2
\end{aligned}
$$

2.3.3   *Probability generating functions*

In engineering, discrete random variables are often used for counting purposes. for example, they can be used to represent the number of circuits busy on a telephone route or the number of broken down machines in a factory. In fact, in most applications, the image space $X(\Omega) = \{x_0, x_1, x_2, \ldots\}$ is a subset of the integers. For the remainder of this section we shall consider only non-negative integer valued discrete random variables.

We now introduce the concept of a probability generating function (PGF), such that it will represent the probabilities of the discrete variable in a single expression. This is achieved by expanding the expression in the form of a power series such that the coefficient of each term of the series represents a probability. Once a suitable generating function has been found to represent a probability distribution, it may then be manipulated to develop some of the properties of the original distribution.

Let X be a non-negative integer-valued random variable with probability distribution $\{p_0, p_1, p_2, \ldots\}$, where $p_i = P(X = i)$ for $i = 0,1,2,\ldots$. For any real number z, the function z^X is itself a discrete random variable. The expectation of z^X defines a function of z,

$$G_X(z) = E(z^X) = \sum_{i=0}^{\infty} p_i z^i$$

$$(2.8)$$

which closely resembles the z-transform of the sequence $\{p_0, p_1, p_2, \ldots\}$. $G_X(z)$ is then the probability generating function of the distribution of X. If the PGF of X is a known function of z, then the probability distribution of X can be calculated by expanding the PGF into a power series about $z=0$, using for example the Taylor series expansion formula. Equivalently, successive differentiation of the PGF with respect to z obtains

$$G_X(0) = p_0$$

$$G_X^{(1)}(0) = \frac{d}{dz} G_X(z) \Big|_{z=0}$$

$$= \sum_{i=0}^{\infty} i\, p_i z^{i-1} \Big|_{z=0} = p_1$$

and for $k \geq 1$,

$$G_X^{(k)}(0) = \frac{d^k}{dz^k} G_X(z) \Big|_{z=0} = k!\, p_k$$

$$(2.9)$$

This relationship underlies the importance of the PGF: in practical problems where it is difficult to calculate a distribution directly, it may be easier to first calculate the PGF, and obtain the distribution from that. Convergence of the infinite series defining the PGF is assured if $0 \leq z \leq 1$. Note that for all PGFs

$$G_X(1) = \sum_{i=0}^{\infty} p_i = 1$$

The PGF may also be used to calculate moments efficiently, again by successive differentiation with respect to z. For example, the first moment of X, i.e. the mean $E(X)$, is given by

$$G_X^{(1)}(1) = \frac{d}{dz} G_X(z) \Big|_{z=1}$$

$$= \sum_{i=1}^{\infty} i\, p_i\, z_{i-1} \Big|_{z=1}$$

$$= \sum_{i=1}^{\infty} i\, p_i = E(X)$$

$$(2.10)$$

A second differentiation obtains

$$G_X^{(2)}(1) = \frac{d^2}{dz^2} G_X(z) \Big|_{z=1}$$

$$= \sum_{i=2}^{\infty} i(i-1)\, p_i\, z^{i-2} \Big|_{z=1}$$

$$= \sum_{i=2}^{\infty} i^2 p_i - \sum_{i=2}^{\infty} i\, p_i = E(X^2) - E(X)$$

Thus, the second moment of X is given by

$$E(X^2) = G_X^{(2)}(1) + G_X^{(1)}(1)$$

and the variance of X by

$$Var(X) = E(X^2) - \{E(X)\}^2$$

$$= G_X^{(2)}(1) + G_X^{(1)}(1) - \{G_X^{(1)}(1)\}^2$$

$$(2.11)$$

2.3.4 *Some discrete probability distributions*

It is expected that the reader will be familiar with the most common discrete probability distributions of integer valued random variables. The most important distributions and their probability generating functions are summarised below. The distributions are illustrated in Figure 2.1

Bernoulli distribution
The Bernoulli distribution may be used to model any system with just two states of

interest, e.g. a switch element (OFF/ON), or a telephone circuit (FREE/BUSY). The system state is represented by a random variable X with

$$P(X = 1) = p \text{ and } P(X = 0) = 1 - p \qquad (0 < p < 1)$$

Thus X is an integer valued random variable with $p_0 = 1-p$, $p_1 = p$, and $p_i = 0$ for $i \geq 2$. In this case the PGF of X is

$$G_X(z) = 1 - p + pz$$

(2.12)

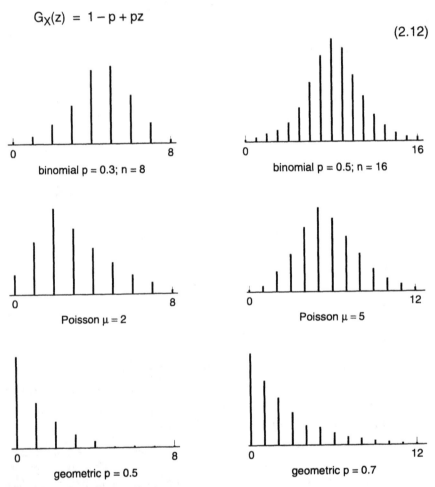

binomial p = 0.3; n = 8

binomial p = 0.5; n = 16

Poisson μ = 2

Poisson μ = 5

geometric p = 0.5

geometric p = 0.7

Figure 2.1 Discrete distributions

Binomial distribution
Consider, for example, a block of n digits sent over a noisy channel in which the probability of corruption of a digit is p. We wish to find the probability that the

block will contain exactly x digits in error. (This is useful in connection with some error correcting codes, which enable the message to be received correctly if 0, 1, . . .x digits are corrupted.)

The probability that a particular x bits are corrupted and the rest correct is

$$p^x (1 - p)^{n-x}$$

Hence the probability that x bits in the block will be corrupted is

$$P(X = x) \;=\; \binom{n}{x} p^x (1-p)^{n-x}$$

As an alternative approach using a generating function, consider the number of errors in one digit position. This is a Bernoulli distribution with generating function

$$G_1(z) \quad = \quad 1 - p + pz$$

The number of errors in n digits is the sum of n such independent variables. It therefore has the generating function

$$G_n(z) \;=\; (1-p-pz)^n$$

$$\;=\; \sum \binom{n}{x} p^x z^x (1-p)^{n-x}$$

$$(2.13)$$

Hence

$$P(X = x) \;=\; \binom{n}{x} p^x (1-p)^{n-x}$$

as before.

Poisson distribution

In the case of the block of digits considered above, suppose the size of the block, n increased indefinitely and let

$$E(X) \quad = \quad np \quad = \quad \mu$$

then

$$G_n(z) \;=\; \left[1 + \frac{\mu(z-1)}{n} \right]^n$$

and in the limit,

$$G_n(z) \;=\; e^{\mu(z-1)} \;=\; e^{-\mu} \sum_{x=0} \frac{(\mu z)^x}{x!}$$

The probability distribution is then

$$P(X=x) \;=\; \frac{\mu^x}{x!}\, e^{-\mu}$$

(2.14)

The Poisson distribution is useful in connection with random events occurring in a continuum of time. For example, consider calls being generated by subscribers on a telephone exchange. The calls may be considered to occur at random. Suppose the probability of a call being generated in the elemental time interval dt is kdt. If the time interval is sufficiently small, the probability of two calls in the same interval becomes negligibly small. Then the expectation of the number of calls during a finite interval, T is given by

$$E(R) \;=\; \frac{kT}{\delta t} \;=\; \mu$$

(2.15)

Then the probability of a specific number of calls x arriving in that interval is given by Equation 2.14.

> *Example: calls arrive at a PBX switchboard at the rate of*
> *200 per hour. What is the probability that the operator will*
> *have to deal with three or more calls during a period of 1*
> *minute.*

The expected number of calls during a 1 minute interval is 200/60 = 3.333

Then $P(R = 0)$ = $e^{-3.333}$ =   0.036

$P(R = 1)$ = $3.333\, e^{-3.333}$ =   0.119

$P(R = 2)$ = $\dfrac{3.333^2}{2!}\, e^{-3.333}$ =   0.198

Hence $P(R \geq 3)$ =   0.647

Geometric distribution

Consider a binary digit stream in which a particular digit is received in error. Suppose the probability of the subsequent digit also being received in error is p and the probability of it being received correctly is (1 − p). This will be familiar to communication engineers as the phenomenon of error 'bursts' when, for example, signals are received over a fading radio link. The duration of an error burst will be determined by a succession of errors followed by a correctly received bit. We are interested in the probability distribution of the lengths of the error bursts.

$$P(X = x) = p^x (1-p)$$

The generating function is

$$G_1^{(1)}(z) = \sum p^x (1-p) z^x = \frac{1-p}{1-pz}$$

(2.16)

so that the probability distribution

$$P(X = x) = p^x (1-p)$$

Differentiation of the generating function gives

$$G_1^{(1)}(z) = \frac{p(1-p)}{(1-pz)^2}$$

and

$$G_1^{(2)}(z) = \frac{2p(1-p)^2}{(1-(1-p)z)^3}$$

Then the mean

$$E(X) = \frac{1-p}{p} = \frac{1}{p} - 1$$

(2.17)

and variance

$$Var(X) = \frac{1-p}{p^2}$$

(2.18)

2.3.5   *Discrete random vectors*

Stochastic models of practical systems often involve two or more random variables defined on the same sample space. For example, a complete description of the state of the telephone route consisting of n circuits considered earlier required n separate variables. It is convenient to think of a collection of random variables on a single sample space as a vector-valued variable, that is a random vector.

If X_1 and X_2 are integer valued random variables on the sample space Ω, then $\mathbf{X} = (X_1, X_2)$ is a random vector of two components on Ω, with image space $\mathbf{X}(\Omega) = \{(i,j) \mid i,j = 0,1,2, \ldots\}$. For example, $\mathbf{X}$ might represent the pair of numbers showing on two dice thrown simultaneously. The probability distribution of $\mathbf{X}$ is the two dimensional sequence $\{p_{ij} \mid i,j = 0,1,2, \ldots\}$ where

$$p_{ij} = P(X_1 = i \ \& \ X_2 = j)$$

This sequence is usually called the joint distribution of X_1 and X_2. As for any discrete probability distribution it must sum to unity.

$$\sum_{i}^{\infty} \sum_{j}^{\infty} p_{ij} = 1$$

Now for $j = 0,1,2, \ldots$ the events $X_2 = j$ are mutually exclusive and exhaustive; hence, by the theorem of total probability

$$P(X = i) = \sum_{j=0}^{\infty} P(X_1 = i \ \& \ X_2 = j) = \sum_{j=0}^{\infty} p_{ij}$$

(2.19)

that is, the marginal distribution of X_1 is obtained from the joint distribution of X_1 and X_2 by summation.

If $\mathbf{X} = (X_1, X_2)$ is a random vector and $\phi : R^2 \to R$ is a real-valued function of ordered pairs of real numbers, then $\phi(X_1, X_2)$ is a discrete random variable. The expectation of $\phi(X_1, X_2)$ is defined to be

$$E[\phi(X_1, X_2)] = \sum_{i=0}^{\infty} \sum_{j=0}^{\infty} \phi(i, j) p_{ij}$$

(2.20)

If we set $\phi(x,y) = x+y$ we obtain an extremely important identity:

$$E(X_1 + X_2) \quad = \quad E(X_1) + E(X_2)$$

(2.21)

This property combined with property (iii) in Section 2.3.2 shows that the expectation operator E is a linear operator on the space of random variables. Thus the expectation of $X_1 + X_2$ may be computed directly from the marginal distributions of X_1 and X_2, without knowledge of their joint distribution. In general, however, the distribution of $X_1 + X_2$ can only be computed from the joint distribution of X_1 and X_2 for $k = 0,1,2, \ldots$.

$$P(X_1 + X_2 = k) \quad = \quad \sum_{i+j=k} p_{ij} \quad = \quad \sum_{i=0}^{k} p_{i,k-i}$$

(2.22)

2.3.6 Independent random variables

Two random variables X_1 and X_2 defined on the same sample space are said to be mutually independent if every conceivable event concerning X_1 only is independent of every conceivable event concerning X_2 only. Now every conceivable event concerning a discrete random variable X comprises a disjoint union of distinct elementary events of the form '$X = x$'. Hence, two integer valued random variables, X_1 and X_2 are independent if and only if

$$P(X_1 = i \ \& \ X_2 = j) \quad = \quad P(X_1 = i) \ P(X_2 = j) \quad \text{for all } i,j = 0,1,2, \ldots$$

In other words, the joint distribution of two independent integer valued random variables factorises into the product of the marginal distributions. Conversely, if it can be shown that two random variables are independent, then their joint probability distribution may be calculated simply from their respective distributions, by straightforward multiplication. In practice, however, the principal significance of independence is that the factorisation property greatly simplifies the analysis of models involving combinations of random variables. In fact, unless independence can be established the solution of many standard problems is intractable. Moreover, approximate solutions to more complex problems often require that independence be assumed to hold even when it is known that it does not.

The summation of random variables offers a striking example of the power of the independence concept. If X_1 and X_2 are two independent random variables, then the distribution of $X_1 + X_2$ previously obtained further simplifies to become

$$P(X_1 + X_2 = k) \quad = \quad \sum_{i=0}^{k} P(X_1 = i) \, P(X_2 = k-i) \qquad \text{for } k = 0,1,2,.....$$

(2.23)

At this stage it is necessary to introduce the concept of convolution. Thus, if X_1 and X_2 are independent, the probability distribution of $X_1 + X_2$ may be calculated by convoluting the discrete probability distribution of X_1 with that of X_2. Note that it is not necessary to first determine the joint distribution of X_1 and X_2. It may be shown that the discrete convolution of two sequences $\{a_n; n = 0,1,2, \ldots\}$ and $\{b_n; n = 0,1,2, \ldots\}$ is another sequence $\{c_n; n = 0,1,2, \ldots\}$ defined by

$$c_k \quad = \quad \sum_{i=0}^{k.} a_i \, b_{k-i} \qquad \text{for } k = 0, 1, 2, \ldots \ldots$$

(2.24)

If this relationship holds we write:

$$\{c_n\} = \{a_n\} * \{b_n\}$$

where the ' * ' sign is used to denote convolution.

Another common situation in which the independence concept proves useful is the multiplication of RVs. If X_1 and X_2 are independent random variables then

$$E(X_1 X_2) \quad = \quad E(X_1) \, E(X_2)$$

as is easily demonstrated by setting $\phi(x,y) = xy$ in the definition of $E(\phi(X_1, X_2))$. Thus the expectation of a product of two independent random variables is the product of their respective means. More generally, setting $\phi(x,y) = \alpha(x)\beta(y)$ where α and β are arbitrary functions, we find

$$E [\alpha(X_1) \, \beta(X_2)] \quad = \quad E [\alpha(X_1)) \, E(\beta(X_2)]$$

(2.25)

if X_1 and X_2 are independent. In particular, this relationship shows that

$$E(X_1 X_2) \quad = \quad E(X_1) \, E(X_2)$$

for any pair of independent discrete random variables whether they are integer valued or not.

If X_1 and X_2 are random variables, then $X_1 + X_2$ is a random variable whose probability generating function may be expressed as

$$G_{X_1 + X_2}(z) \;=\; E(z^{X_1 + X_2}) \;=\; E(z^{X_1} . z^{X_2})$$

This may be simplified by setting $\alpha(x) = \beta(x) = z^X$ in $E\,[\alpha(X_1)\,\beta(X_2)]$ to obtain

$$G_{X_1 + X_2}(z) \;=\; E(z^{X_1})\,E(z^{X_2}) \;=\; G_{X_1}(z)\,G_{X_2}(z)$$

$$(2.26)$$

Thus the probability generating function of a sum of two independent random variables is the product of the two respective probability generating functions. This result offers an alternative to the convolution method of calculating the distribution of $X_1 + X_2$. First compute the PGFs of X_1 and X_2, then multiply them together to obtain the PGF of $X_1 + X_2$, and expand this into a power series to obtain

$$P(X_1 + X_2 = k) \quad \text{for } k \;=\; 0, 1, 2, \ldots.$$

in the usual way. This method often requires considerably less calculation than the direct convolution of the distributions of X_1 and X_2.

The concept of independence is readily extended to an arbitrary number of random variables on the same sample space. The n random variables $X_1, X_2, \ldots, X_n$ are said to be mutually independent if

$$P\left(\underset{i=1}{\overset{n}{\&}}\,(X_i = x_i) \right) \;=\; \prod_{i=1}^{n} P(X_i = x_i)$$

$$(2.27)$$

for all integers $x_1, x_2, \ldots, \quad (x \;=\; 0,1,2,\ldots)$

If $X_1, X_2, \ldots, X_n$ are independent random variables, their sum $Y_n = X_1 + X_2 + \ldots + X_n$ is a random variable whose PGF is the product of the PGFs of the X_i.

$$G_{Y_n}(z) \;=\; \prod_{i=1}^{n} G_{X_i}(z)$$

$$(2.28)$$

This follows because

$$E(z^{Y_n}) = E(z^{X_1 + X_2 + \ldots X_n}) = E(z^{X_1} \cdot z^{X_2} \ldots \ldots z^{X_n})$$
$$= E(z^{X_1}) \cdot E(z^{X_2}) \ldots \ldots E(z^{X_n})$$

since the X_i are independent.

> *Example: Consider a sequence of n bits. Suppose each bit is assigned the value 0 or 1 with probability q and p, respectively. What is the distribution of the total number of 1's in the sequence?*

Let X_i be the value assigned to the ith bit. Then each X_i has a Bernoulli distribution : $P(X_i = 0) = q$, $P(X_i = 1) = p$.

The probability generating function is $q + pz$.

The total value of the sequence is $Y_n = X_1 + X_2 + \ldots \ldots + X_n$, whose probability generating function is the product of the PGFs of the independent X_i

$$G_{Y_n}(z) = (q + pz)^n = \sum_{k=0}^{n} \binom{n}{k} q^{n-k}(pz)^k$$

The PGF is easily expanded into a finite series using the binomial theorem.

Thus Y_n has the binomial distribution

$$P(Y_n = k) = \binom{n}{k}(1 - p)^{n-k} p^k \qquad \text{for } k = 0,1,2, \ldots ,n$$

2.3.7   Compound randomness

Suppose $X_1, X_2, X_3, \ldots$ form an infinite sequence of independent, identically distributed (IID) random variables and let N be another independent random variable. Then the sum $S = X_1 + X_2 + \ldots + X_N$ has a random number of terms, i.e. it is a compound random variable. Compound randomness arises naturally in various problems of communications engineering. For instance, N might represent the number of bytes of data in a message, and X_i the number of bit errors received in the ith byte, in which case S would correspond to the total number of errors in the message. We can determine the distribution of S in the general case by means of appropriate conditioning on N.

The event 'N = n' partitions the sample space into an infinite number of mutually exclusive and exhaustive subsets distinguished by the different possible values of n, i.e. 0,1,2,3, . . . Hence, we may apply the theorem of total probability to the event 'S = k' to obtain

$$P(S = k) = \sum_{n=0}^{\infty} P(S = k \& N = n) = \sum_{n=0}^{\infty} P(S = k \mid N = n) \, P(N = n)$$

$$= \sum_{n=0}^{\infty} P(X_1 + X_2 + \ldots + X_n = k) \, P(N = n)$$

$$(2.29)$$

where the conditioning event 'N = n' may be omitted because N is independent of the X_i. This expansion of $P(S = k)$ obtains the PGF of S

$$G_s(z) = \sum_{k=0}^{\infty} P(N = n) \, z^k$$

$$= \sum_{k=0}^{\infty} \sum_{n=0}^{\infty} P(X_1 + X_2 + \ldots + X_n = k) \, P(N = n) \, z^k$$

$$= \sum_{n=0}^{\infty} P(N = n) \sum_{k=0}^{\infty} P(X_1 + X_2 + \ldots + X_n = k) \, z^k$$

$$= \sum_{n=0}^{\infty} P(N = n) \, G_{X_1 + \ldots + X_n}(z)$$

$$= \sum_{n=0}^{\infty} P(N = n) \, \{G_X(z)\}^n$$

$$= G_n\{G_X(z)\}$$

$$(2.30)$$

Example. If the number of bytes per message has PGF

$$G_N(z) = \frac{1-a}{1-az}z \qquad \text{where } 0 < a < 1 \text{ is constant}$$

and the number of errors per byte has a binomial distribution such that $G_X(z) = (1 - p + pz)^8$, *find the PGF for the total number of errors per message and show that the expected number of errors per message is equal to the expected number of bytes per message times the expected number of errors per byte.*

$$
\begin{aligned}
G_s(z) &= G_N(G_X(z)) \\
&= \frac{(1-a)(1-p+pz)^8}{1-a(1-p-pz)}
\end{aligned}
$$

Then expected number of errors per message,

$$E(S) = G'_s(1) = \frac{8p}{1-a}$$

Expected number of bytes per message

$$E(N) = G'_N(1) = \frac{1}{1-a}$$

Expected number of errors per byte

$$E(X) = G_X'(1) = 8p$$

which proves the relationship.

2.4 Continuous random variables

2.4.1 *Distribution function*

There are many stochastic systems in engineering in which the behaviour of interest is not readily reduced to a discrete random variable. For example, the duration of a telephone call, or the lifetime of an integrated circuit, may assume any finite positive value. In order to model such quantities, the image set of the random variable must be extended to include the continuous semi-infinite interval $(0,\infty)$, which contains an infinite number of distinct points. However, this takes the model beyond the scope of discrete probability theory since, if a probability $p(\omega_i > 0)$ is ascribed to each point, then the sum of the probabilities would be infinite whereas we know that the sum of the probabilities must be equal to 1. To get around this problem, when dealing with a continuous random variable or sample space we do not assign probability to individual sample points, but rather to intervals on the real line. Thus $P(\omega\varepsilon\ [a,b]) \geq 0$, but $P(\omega_i) = 0$ for any ω_i. Clearly, the most elementary event of interest concerning a continuous random variable X cannot be 'X = x' as it was for discrete random variables. Instead we take 'X ≤ x' = $\{\omega\varepsilon\Omega \mid X(\omega) \leq x\ \}$ to be the basic event.

The probability of the basic event 'X ≤ x' defines a function of $x\varepsilon R$ which is of central importance in the study of all random variables, whether they be discrete or continuous or a mixture of the two. For any random variable X we define the distribution function (DF) of X to be $F_X:R \rightarrow [0,1]$ such that

$$F_x(x)\ =\ P(X \leq x) \qquad \text{for each}\ \ x \in R$$

(Note that the subscript x is merely a label to distinguish F_x from F_y; the argument x stands for any real number, i.e. $F_x(x)$ is a function of x.)

Given the distribution function of X, the probability of any event may be calculated. For example, for any $a < b$

$$P(a < X \leq b)\ =\ P(X \leq b) - P(X \leq a)\ =\ F_x(b) - F_x(a).$$

The properties of the distribution function are:

(i) F_X is a nondecreasing function: $x < y \Rightarrow F_X(x) \leq F(y)$
(ii) $\lim_{x \rightarrow +\infty} F_X(x)\ =\ 1$
(iii) $\lim_{x \rightarrow -\infty} F_X(x)\ =\ 0$

A continuous random variable has a continuous distribution function. A discrete random variable has a staircase distribution function. A random variable which is neither purely continuous nor purely discrete is said to be of mixed type and has a distribution function which has a finite number of (positive jump) discontinuity points, in between which it is continuous.

A continuous random variable which has a differentiable distribution function $F_X(x)$ is said to be absolutely continuous. The derivative $f_X(x) = F_X'(x)$ is called the probability density function (PDF) of X.

The properties of the probability density function are:

(i) $f_X(x) \geq 0$ for all $x\varepsilon R$

(ii) $\int_{-\infty}^{\infty} f_X(x)\, dx = 1$

(iii) $P(a < X < b) = \int_a^b f_X(x)\, dx$ for $a < b$

Thus the probability that X takes a value belonging to any interval [a,b] of the real line is represented by the area under the curve $y = f_X(x)$ between x=a and x=b. These properties follow from those of the distribution function and the fundamental theorem of calculus:

$$F_X(x) = \int_{-\infty}^{x} f_X(t)\, dt \qquad \text{for} \qquad -\infty < x < \infty$$

2.4.2   *Expectation*

If X is a continuous RV, with PDF $f_X(x)$, and $\phi:R \to R$ is an arbitrary function, we define the expectation of f(X) to be

$$E(\phi(X)) = \int_{-\infty}^{\infty} \phi(x) f_X(x)\, dx$$

It is readily shown that this expectation operator for continuous random variables possesses the properties already established for the discrete operator in Section 2.3.2.

As for discrete random variables, the mean and nth moment of X are defined as E(X) and $E(X^n)$, respectively, and the variance of X is

$$Var(X) = E((X - E(X))^2) = E(X^2) - (E(X))^2$$

For any random variable, X the coefficient of variation C_X is defined to be

$$C_X = \frac{\sqrt{Var(x)}}{E(X)}$$

(2.31)

2.4.3   The Laplace transform

The expectation of exp(-sX), where s is an arbitrary real number, defines a real function of s

$$L_X(s) = E(e^{-sX}) = \int_{-\infty}^{\infty} e^{-sX} f_X(x)\ dx$$

(2.32)

which is readily identified as the two-sided Laplace transform of the probability distribution function of X, $f_X(x)$. The integral is well-defined for $s \geq 0$, since then $0 \leq L_X(s) \leq L_X(0) = 1$. Note that the lower limit of the integral must be $-\infty$, and not zero, because in general random variables may assume negative as well as positive values. However, if $F_X(0) = P(X \leq 0) = 0$, then $L_X(s)$ is the same as the familiar one-sided Laplace transform where the lower limit of integration is zero. Since there is a one-to-one relationship between any function and its Laplace transform the function $L_X(s)$ offers a complete characterisation of the random variable X: if $L_X(s)$ is given, then to obtain the PDF and hence the distribution function, of X all we have to do is find the inverse of $L_X(s)$. Thus the Laplace transform performs a role for continuous random variables similar to that performed by the probability generating function for integer valued random variables.

The calculation of moments from the Laplace transform is even more straightforward than using the PGF. Substituting the power series representation of exp(-sX) into the definition of $L_X(s)$ we find

$$L_X(s) = E\left(\sum_{k=0}^{\infty} \frac{(-sX)^k}{k!}\right) = \sum_{k=0}^{\infty} \frac{(-s)^k}{k!} E(X^k)$$

(2.33)

after making repeated use of the expectation operator properties given in Section 2.3.2. Thus the nth moment of X may be derived from the power series expansion of $L_X(s)$ around $s = 0$ by multiplying the coefficient of s^n by $(-1)^n n!$. Alternatively, successive differentiation of the Laplace transform with respect to s may be used to calculate $E(X^n)$, because

$$L_X^{(n)}(0) = \frac{d^n}{ds^n} L_X(s) \qquad |s = 0$$

$$= \int_{-\infty}^{\infty} (-x)^n e^{-sx} f_X(x) \, dx \qquad |s = 0$$

$$= (-1)^n E(X^n) \tag{2.34}$$

where we have differentiated through the integral sign.

2.4.4   *Some continuous probability distributions*

It is expected that the reader will have some knowledge of the most common continuous probability distributions. The most important ones are summarised below and the distributions are illustrated in Figure 2.2.

Uniform (or rectangular) distribution
It follows from previous discussion that a quantity which can take any value from a to b with uniform probability has the density function

$$f(x) = \frac{1}{b-a} \qquad \text{if } a < x < b$$

$$\qquad\qquad 0 \qquad\qquad \text{otherwise}$$

$$\tag{2.35}$$

The mean is evidently $(a + b)/2$ and the variance may be shown to be

$$V(x) = \frac{(b-a)^2}{12}$$

Normal or Gaussian distribution
Many natural phenomena such as Brownian motion (displacement) and random noise (amplitude) may be described in terms of the normal distribution. The probability density is given by

$$f(x) = \frac{1}{\sqrt{2\pi}\ \sigma} \exp\left[\frac{-(x-\mu)^2}{2\sigma^2}\right]$$

$$(2.36)$$

The mean is μ and the variance σ^2.

Even if the variable itself does not follow a normal distribution, some simple function of it may do so. A common example is the use of a logarithmic or decibel scale to measure signal strength or a 'pH' value to measure acidity/alkalinity in chemistry. The distribution is then known as log-normal and takes the form given in Equation 2.36 with 'log x' substituted for 'x'.

Gamma distribution
A continuous random variable whose density is given by

$$f(x) = \frac{\lambda^r}{\Gamma(r)} x^{r-1} e^{-\lambda x}$$

$$(2.37)$$

is said to be a gamma random variable where the gamma function is defined by

$$\Gamma(r) = \int_0^{\infty} e^{-y} y^{r-1} dy$$

2.4.5   Exponential distribution

The exponential distribution is particularly important in performance engineering and so its properties will be considered in somewhat greater depth. In many examples of service times such as transactions at a bank counter or telephone calls, it is often found that the distribution of the durations is well represented by an exponential distribution.

The probability density function takes the form

$$f(x) = \mu e^{-\mu x}$$

$$(2.38)$$

This distribution has the useful property that the mean is $1/\mu$ and the variance $1/\mu^2$.

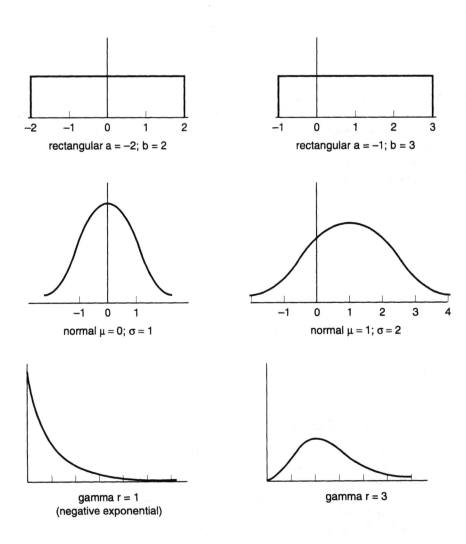

Figure 2.2   Continuous distributions

The value of the parameter μ may be adjusted to match $E(X)$ to the average service time of the system.

The probability that the service time will be less than a given value is given by the cumulative distribution

$$P(X \le x) \;=\; 1 - e^{-\mu x} \qquad\qquad \text{for } 0 \le x \le \infty$$

$$(2.39)$$

Consider, for example, a telephone call which was set up at time zero and is still in progress at time t (Figure 2.3). For how much longer is the call likely to continue before finally clearing down?

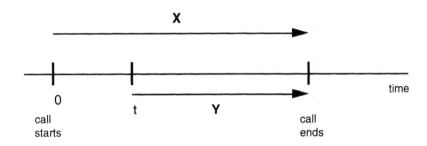

Figure 2.3   Timing for telephone call distribution

Define X = the total duration of the call,
   and   Y = the residual duration of the call at time t.
Then   Y = X−t is the future time remaining before the call ends.

We seek the conditional distribution of the random variable Y, given that X > t.

Now

$$P(Y > u \mid X > t) \;=\; P(X - t > u \mid X > t) \;=\; P(X > t + u \mid X > t)$$

$$=\; \frac{P(X > t + u \ \& \ X > t)}{P(X > t)} \;=\; \frac{P(X > t + u)}{P(x > t)}$$

$$=\; \frac{e^{-\mu(t+u)}}{e^{-\mu t}} \;=\; e^{-\mu u} \qquad\qquad \text{for all } u > 0$$

Hence

$$P(Y \le u \mid X{>}t) \;=\; 1 - e^{-\mu u}, \qquad 0 \le u < \infty.$$

Thus, however long a call has already been in progress (i.e. time t), the conditional distribution of the residual duration of the call is the same (i.e. it is independent of t), and moreover it is the same as the exponential distribution of the duration of a complete telephone call! In other words, a consequence of assuming that call duration is exponentially distributed is that the telephone users do not 'remember' how long they have been already been talking.

This memoryless property is a feature unique to the exponential distribution, which is often used in stochastic models because it considerably simplifies the subsequent analysis.

2.5   Random point processes

2.5.1   Stochastic modelling of random events

In order to analyse the behaviour of any system involving random events, we first require a model that determines the time instants at which the events occur. In most cases the generation of the events is as a result of complex social and commercial interactions such as people arriving at a bus queue or the generation of telephone calls. It may not be necessary for human agencies to be involved. An example of a mechanistic generation of random events would be a case in which the event was triggered by a noise waveform exceeding a predetermined level. The result is that it is impossible to predict the incidence of events over some future period of time. In the absence of a deterministic model for this complex process, we should like to construct a stochastic model. However, to bring this situation within the framework of probability theory, we have to conceive of an underlying random experiment, each of whose elementary outcomes is a particular set of discrete points on the infinitely long real line. Such a random experiment defines a random point process, which is one kind of stochastic process.

It would clearly be impractical to study a point process described only by patterns of points on a line. Instead, we introduce a function of time $A(t)$ which is a count of the number of point events which occur in the time interval $(0,t)$, measured from some convenient time origin. For each $t > 0$, $A(t)$ is an independently valued random variable, with $A(0)=0$. For any particular realisation of the random point process, $\{A(t); 0 \leq t < \infty\}$ is a staircase function gradually stepping up through the positive integers. Clearly, each realisation uniquely defines a staircase function on $(0,\infty)$, and vice versa. In the context of the random experiment, however, $\{A(t); 0 \leq t < \infty\}$ is a random counting process, an example of a discrete-valued, continuous parameter stochastic process, i.e. an infinite collection of integer valued random variables indexed by the parameter time.

2.5.2   The Poisson process

The Poisson process is defined as the random point process, whose associated counting process satisfies the following conditions:

(i) $A(0) = 0$

(ii) If $0 < t_1 < t_2 < t_3 < t_4$ then $A(t_2) - A(t_1)$ and $A(t_4) - A(t_3)$ are independent random variables, i.e. $\{A(t); 0 \le t < \infty\}$ has 'independent increments'.

(iii) The distribution of $A(t+s) - A(t)$ is independent of t, and is the same as that of $A(s)$, i.e. $\{A(t); 0 \le t < \infty\}$ has 'stationary increments'.

(iv) P[one point event in $(t,t+h)$] $= \lambda h + o(h)$

(v) P[two or more point events in $(t,t+h)$] $\rightarrow o(h)$

where for any function f we write ' $f(h) = o(h)$ ' to indicate that as $h \rightarrow 0$ the function $f(h)$ tends to zero faster than h itself, that is $\lim_{h \rightarrow 0} f(h)/h = 0$.
In other words, the effect of the term $o(h)$ in (iv) and (v) is negligible if h is very small.

Thus, in a Poisson process point events occur singly and at a rate λ which is uniform in time. In addition, the pattern of points in a given time interval is independent of the pattern in any other disjoint, i.e. non-overlapping, interval, e.g. call-attempts in the interval (s,t) are independent of call-attempts in $(0,s)$ and (t,∞). Hence, the Poisson process is sometimes called the completely random point process, and we say that the point events belonging to a Poisson process occur at random.

From the definition of a Poisson process, several important properties may be deduced.

A(t) has a Poisson distribution
The number of arrivals in any time interval of length t has a Poisson probability distribution with mean λt:

$$P[A(t) = n] = e^{-\lambda t} \frac{(\lambda t)^n}{n!} \qquad (n = 0, 1, 2, \dots)$$

(2.40)

To prove this relationship, denote the discrete probability distribution of $A(t)$ by $p_n(t) = P(A(t)=n)$ to emphasise its dependence on time. The PGF of $A(t)$ defines a function of both z and t:

$$P(z,t) \;=\; E\,(z^{A(t)}) \;=\; \sum_{n=0}^{\infty} P_n\,(t)\; z^n$$

$A(0) = 0$ implies that $P(z,0) = 1$.

Consider the adjacent time intervals $(0,t)$ and $(t,t+h)$, where h is small. If the integer valued random variable, $Y(h)$ denotes the number of events which occur in $(t,t+h)$, then

$$P(Y(h) = 0\,) \;=\; 1 - \lambda h + o(h)$$
$$P(Y(h) = 1\,) \;=\; \lambda h + o(h)$$
and $\qquad P(Y(h) \geq 2\,) \;=\; o(h)$

Hence the PGF of $Y(h)$ is

$$E(\, z^{Y(h)}\,) \;=\; 1 - \lambda h + \lambda hz + o(h).$$

Now
$$A(t+h) \;=\; A(t) \;+\; Y(h),$$

and because (ii) implies that $A(t)$ and $Y(h)$ are independent random variables, the PGF of their sum is equal to the product of their PGFs. Hence,

$$P(z,t+h) \;=\; P(z,t)\,\{\, 1 - \lambda h + \lambda hz + o(h)\,\}$$

We may rearrange this equation to obtain

$$\frac{P(z,\, t+h)\;-\;P(z,\,t)}{h} \;=\; P(z,\,t)\,[\lambda\,(z-1)\;+\;\frac{o(h)}{h}\,]$$

Letting h tend to zero, it follows that

$$\frac{dP(z,\,t)}{dt} \;=\; \lambda\,(z-1)\,P(z,\,t) \qquad\qquad 0 \leq t < \infty$$

Hence

$$\frac{d}{dt}\, \log P(z,\,t) \;=\; \frac{1}{P(z,\,t)}\,\frac{dP(z,\,t)}{dt} \;=\; \lambda\,(z-1)$$

Integrating this equation on (0,t) shows that

$$\log P(z,t) - \log P(z,0) = \lambda t (z-1)$$

Hence,

$$P(z, t) = e^{\lambda t(z-1)}$$

which is recognisable as the PGF of a standard Poisson distribution with parameter λt. Thus, $A(t)$ has the Poisson distribution as claimed above. Moreover, (iii) implies that $A(s+t) - A(s)$ has the same distribution as $A(t)$. Hence, the number of events in any interval of length t has a Poisson distribution with mean λt.

Since $E(A(t)) = \lambda t$, the mean number of events occurring in any time interval is proportional to the length of the interval. Thus, the parameter λ has a physical interpretation as the average rate of occurrence of point events.

The interevent time distribution is exponential

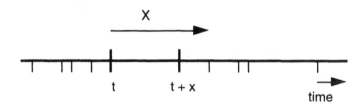

Figure 2.4 Event timing

At any given time t, the lapse of time X until the occurrence of the next subsequent event (at time $t+X$) is a random variable . Now $X>x$ if and only if there are no events in the interval $(t,t+X)$ (Figure 2.4). Therefore,

$$P(X>x) = P(A(t+x) - A(t) = 0) = P(A(x)=0) = e^{-\lambda x}$$

since the Poisson process is stationary, and $A(x)$ has a Poisson distribution of mean λx. Hence, the random variable X has an exponential distribution

$$P(X \leq x) = 1 - e^{-\lambda x} \qquad\qquad x \geq 0,$$

$$(2.41)$$

which does not depend on the given time t, nor on the previous history of the process prior to time t. Thus, the Poisson process is a memoryless stochastic point process.

Property (ii) implies that the above reasoning remains valid even if the time instant t coincides with a point event. Hence, the durations of the periods of time between successive events are exponentially distributed with mean $1/\lambda$.

2.6   Correlation and covariance

2.6.1   Cross correlation

In Section 2.3.6, we considered the conditions for independent random variables. However, it is important to consider the effects of variables that are not independent. In this case, the variables are said to be correlated and a measure of the extent of the interdependence is the covariance. For example, the midday temperatures in London and Paris are correlated since both cities are in the northern hemisphere. Consider two variables X_1 and X_2 and suppose that there is a tendency for departures from their respective means to be mutually dependent. The covariance is defined as

$$Cov(X_1, X_2) = E\{[X_1 - E(X_1)][X_2 - E(X_2)]\}$$

$$(2.42)$$

It may be shown that

$$Cov(X_1, X_2) = E(X_1 X_2) - E(X_1) E(X_2)$$

and $$Var(X_1 + X_2) = V(X_1) + V(X_2) + 2\, Cov(X_1 X_2)$$

A dimensionless measure of the extent of the covariance is the correlation coefficient, defined as

$$\rho(X_1, X_2) = \frac{Cov(X_1, X_2)}{[V(X_1)\, V(X_2)]^{1/2}}$$

$$(2.43)$$

The correlation coefficient takes values between +1 and -1; a zero value indicates that X_1 and X_2 are independent.

2.6.2 Autocorrelation and autocovariance

If a variable takes successive values X_1, X_2, X_3....... then successive values may exhibit a degree of inter-dependence. For example, the midday temperature in London on 2 March is often (but not always) closer to that measured on 1 March than is the temperature on 31 March. Similarly, in the case of a continuous function that varies slowly with time or distance, two samples taken close together are more closely related than samples taken far apart. Just as we referred to the correlation between two separate variables, so we can talk of the autocorrelation of a variable with itself.

Consider a continuously varying function of time, f(t) such as shown in Figure 2.5(a). Now displace the waveform by a fixed amount, τ. Then the autocovariance is defined as a function of τ by

$$R(t) \quad = \quad E \left[\, (f(t) - \mu) \, . \, (f(t + \tau) - \mu) \, \right]$$

$$(2.44)$$

where μ is the mean value

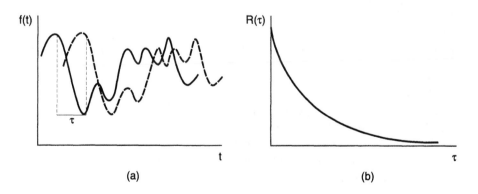

$$(a) \qquad\qquad\qquad (b)$$

Figure 2.5 Derivation of autocovariance function

If the displacement τ is zero, it is evident that the covariance is equal to the variance of the function. For small displacements, if the value of the sample of the waveform at instant t is positive, then the value of the sample at instant $(t + \tau)$ is more likely to be positive than negative. The mean value of the product of a large number of such samples will then be high. However, for small displacements, the values of the two samples will be independently positive and negative with equal probability, so that the covariance will tend to zero. It follows that the autocovariance decays continuously to zero as the time interval is increased as shown in Figure 2.5b.

The autocovariance is a measure of the rate at which a random function can change with the independent variable such as time or distance. For example, random (Gaussian) noise that has been passed through a narrow bandwidth filter has an autocovariance function that decays slowly with the time interval. If the bandwidth of the filter is increased, the autocovariance function decays more rapidly.

Note that autocorrelation, as well as being an attribute, is sometimes given a quantitative meaning

$$R_1(t) \quad = \quad E[f(t) \cdot f(t + t)]$$

(2.45)

This implies that the absolute values of the samples are used rather than the excursions from the mean value.

2.7 Birth and death processes

In the study of biological populations, random point processes are used to model the birth of new individuals. Let $N(t)$ denote the total size of the population at time t. The Poisson process would represent a population in which no deaths occur and the probability of a birth in $(t,t+h)$ is independent of the current size of the population $N(t) = A(t)$, the number of previous births. For some systems, however, it is possible that this probability would depend on $N(t)$. It is also likely that deaths would sometimes occur. The death of an individual would cause $N(t)$ to decrease by one. In this case, each realisation of $N(t)$ is not a staircase function like $A(t)$, but a step function stepping both up and down the nonnegative integers. The simplest stochastic model of this type of situation is called a birth and death process (BDP), but it finds widespread application in many areas of science, including reliability and traffic engineering.

2.7.1 Definition of a birth and death process

The non-negative integer-valued random process $\{N(t); t \geq 0\}$ is a birth and death process (BDP) if the following three conditions are satisfied, for some set of nonnegative constants $\{b_n, d_n; n=0,1,2,...\}$, as $h \to 0$,

(i) $P[\ N(t+h) = n+1 \mid N(t)=n] = b_n h + o(h), \quad n = 0,1,2,...$

(ii) $P[\ N(t+h) = n-1 \mid N(t)=n] = d_n h + o(h), \quad n = 1,2,3,...$

(iii) $P[\ (N(t+h) - N(t)) \geq 2\ ] = o(h).$

$N(t)$ may be thought of as the size of a population at time t, which evolves through time with births and deaths occurring singly so that $N(t)$ can only change by ± 1 at any time instant. Most of the time, however, there will be no change in $N(t)$, since it follows from the above conditions that, as $h \to 0$,

(iv) $P[\ N(t+h) = n \mid N(t) = n\ ] = 1 - (b_n + d_n)h + o(h).$

It is clear that the birth rate and death rate at time t may depend on the value of $N(t)$; however, the values which each may assume are drawn from the sets of birth coefficients $\{b_n; \ n = 0,1,2,..\}$ and death coefficients $\{d_n; \ n = 1,2,3,...\}$, respectively, which are constant for all time. Since $N(t) \geq 0$ we set $d_0 = 0$.

A stochastic process $\{X(t); t \geq 0\}$ is known as a Markov process if its probabilistic behaviour over the time interval (t, ∞) is determined entirely by the state of the system at time t only, that is its future is independent of its past history prior to t. Thus, a birth and death process is a special kind of continuous-time Markov process with discrete state space $\{0,1,2, \ldots \}$, and instantaneous transitions possible only between adjacent states, i.e. from n to n+1 (a birth) or (if n≥1) to n-1 (a death).

2.7.2 *Behaviour of a birth and death process*

A transition diagram (Figure 2.6) is a useful pictorial representation of a birth and death process, showing the state space and indicating the possible state transitions by arrows, labelled by the associated birth and death rates.

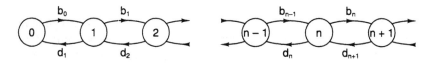

Figure 2.6   Transition diagram representation of a birth and death process

We know that, as time goes on, $N(t)$ steps up and down the state space occasionally, but to complete our understanding of how the process behaves, we need to examine how long it remains in each state before making a transition. Suppose that the birth and death process enters state n (≥ 1) at time t_0. Assuming that no birth or death has occurred by time t_0+t ($t > 0$), the probability of a birth in (t_0+t, t_0+t+h) is analogous to

(i)   the conditional probability that a device with constant failure rate b_n switched on at time t_0 and still functioning at time t_0+t fails in (t_0+t, t_0+t+h),

or   (ii)   the probability that a Poisson process of rate b_n switched on at time t_0 experiences its first event in (t_0+t, t_0+t+h).

Using either analogy, we may deduce that the earliest time $t_0 + X_n$ after t_0 at which a birth would occur, assuming a death does not occur first, defines a random variable with the exponential distribution

$$P(X_n < t) \;=\; 1 - e^{-b_n t}$$

Similarly, the lapse of time Y_n before the first death after t_0 would occur, assuming that a birth does not occur first, has the exponential distribution

$$P(Y_n \leq t) \;=\; 1 - e^{-d_n t} \qquad t \geq 0$$

However, only one of the two possible events can happen first. If $X_n < Y_n$ in this particular realisation of the BDP, there is a birth and the process changes to state $n+1$. If $X_n > Y_n$, there is a death and the process changes to state $n-1$. Thus, the actual time spent in state n at each visit is the minimum of X_n and Y_n, which has the exponential distribution.

$$P(\min \{X_n, Y_n\} \leq t) \;=\; 1 - e^{-(b_n + d_n) t} \qquad t \geq 0$$

As soon as the BDP enters the next state, $n+1$ birth and death rates change to the appropriate values, and the fact that the most recent event was a birth (and, therefore, a time X_n has elapsed since the most recent death) does not affect X_{n+1} and Y_{n+1} because each random variable has a distribution which is memoryless.

2.7.3 Equilibrium distribution of N(t)

Let us define

$$p_n(t) = P(N(t)=n) \qquad \text{for } n=0,1,2,$$

(2.46)

It can be shown, under quite general conditions, that if the process is allowed to carry on for a very long time indeed, then the discrete probability distribution $\{P_n(t)\}$ of N(t) will approach a steady-state or equilibrium distribution $\{P_n\}$ which is independent both of t and of the initial state of the process. Moreover, the equilibrium probability is equal to the proportion of time that the BDP spends in state n over a very long period of time. We shall assume this fact without proof, but we shall deduce what the steady-state distribution must be, and indicate the general conditions necessary for equilibrium to be attainable.

Consider the adjacent time intervals (0,t) and (t,t+h), where h is small. By the theorem of total probability, we may split up $P_n(t+h)$ by conditioning the event 'N(t+h)=n' on all possible distinct values of N(t),

$$p_n(t+h) = \sum_{i=0} P(N(t+h) = n \mid N(t) = i) \, P(N(t) = i)$$

However, for a birth and death process, all but three of these transitions have probability o(h) as $h \to 0$. Hence, for $n \geq 1$,

$$p_n(t+h) = b_{n-1}h \, p_{n-1}(t) + (1 - (b_n + d_n)h) \, p_n(t) + d_{n+1}h \, p_{n+1}(t) + o(h)$$

$$= p_n(t) - h\{(b_n + d_n) \, p_n(t) - b_{n-1} \, p_{n-1}(t) - d_{n+1} \, p_{n+1}(t)\} + o(h)$$

Since we seek only the steady state distribution of N(t), which is independent of t, we now let t tend to infinity in the above equation, thereby showing that it also holds true when $p_n(t+h) = p_n(t) = p_n$. Hence, at equilibrium the coefficient of h must be 0, that is

$$(b_n + d_n) \, p_n = b_{n-1} \, p_{n-1} + d_{n+1} \, p_{n+1} \qquad (n = 1, 2, \ldots)$$

For $n = 0$, some terms are missing because 0 is a boundary state, and we find

$$b_0 \, p_0 = d_1 \, p_1$$

(2.47)

These equations may be recalled easily by referring to the diagram, and noting that the term on the left is the sum of the rates on transitions out of state n, multiplied by P_n, and that the term on the right is the sum of (transition rate) $\times$ (state probability) for transitions into state n. This observation has a physical interpretation in terms of the probability flow between adjacent states. The flow of probability out of state n during $(t,t+h)$ is simply the probability that the BDP is in state n at time t, and leaves it before time $t+h$, that is

$$P(N(t) = n \ \& \ N(t+h) \neq n) = P(N(t+h) \neq n \mid N(t)=n) \ P(N(t)=n)$$

$$= \{ b_n h + d_n h + o(h) \} \ p_n(t).$$

Hence, the rate of probability flow out of state n at time t is

$$\lim_{h \to 0} \frac{\{b_n h + d_n h + o(h)\} \ p_n(t)}{h} = (b_n + d_n) \ p_n(t)$$

$$(2.48)$$

Thus, $(b_n + d_n) \ p_n$ is the rate of probability flow out of state n at equilibrium. Similarly, it can be shown that $b_{n-1} \ p_{n-1}$ is the rate of flow of probability from state $n-1$ into state n at equilibrium, and $d_{n+1} \ p_{n+1}$ is the rate of flow of probability from state $n+1$ into state n at equilibrium.

Thus, the equilibrium equations satisfied by $\{P_n\}$ express the fact that the rate of flow of probability out of state n is exactly balanced by the rate of flow of probability into state n, i.e. the net flow in and out of each state is zero. For this reason the equilibrium equations are usually called balance equations.

Rearranging the balance equation, we find that

$$d_{n+1} \ p_{n+1} - b_n \ p_n = d_n \ p_n - b_{n-1} \ p_{n-1} = \ldots = d_1 p_1 - b_0 p_0,$$

by induction on n.

But $\qquad\qquad d_1 p_1 - b_0 p_0 = 0$

and hence

$$d_n \ p_n = b_{n-1} \ p_{n-1}$$

$$(2.49)$$

This is a simpler form of balance equation which shows that the net probability flow between two adjacent states is also zero. It follows that, provided $n \geq 1$,

$$p_n = \frac{b_{n-1}}{d_n} p_{n-1}$$

and hence, by induction on n,

$$p_n = D_n p_0 \qquad\qquad (n = 0, 1, 2, \dots)$$

where we define $D_0 = 1$, and for $n \geq 1$,

$$D_n = \frac{b_0 b_1 \dots b_{n-1}}{d_1 d_2 \dots d_n}$$

Thus, we have obtained the steady state probabilities $\{p_n\}$ in terms of the transition rates $\{b_n\}$ and $\{d_n\}$, and p_0, which itself must be chosen to normalise the distribution, i.e. to make $\Sigma p_n = 1$. Therefore, we define

$$D = \sum_{j=0}^{\infty} D_j$$

and provided that $D < \infty$, the equilibrium distribution of $N(t)$ is given by

$$P[N(t) = n] = \frac{D_n}{D} \qquad\qquad (n = 0, 1, 2, \dots)$$

$$(2.50)$$

We shall develop the Markov process further in relation to specific applications in subsequent chapters.

Chapter 3

Simulation methods

3.1 Why simulate?

There are a number of approaches that can be used to obtain performance measures from systems and networks. One is to actually measure the performance of the system, but, unfortunately, one cannot ask 20 million telephone subscribers in the UK not to use their telephones one (or more?) day a year to allow experiments to be carried out! Another possibility is to use mathematical analysis, either by hand or on computer, to determine the performance measures of interest. The third is to simulate the system using a computer model.

In this chapter, simulation will refer to the use of a digital computer program that attempts to duplicate the 'real' system. The computer simulation is driven with random inputs supplied by a (pseudo) random-number generator; and the results produced by the simulation are thus stochastic in nature.

Simulation is a valid alternative to analysis for the performance evaluation of systems [1,2], it allows an arbitrary level of detail to be included in the model, and can obtain results for systems for which no tractable analytical model exists. In addition, simulation can be used to model large networks and other systems that are either too extensive, expensive, or both to allow their use for performance experiments (or which simply do not exist yet!).

The major disadvantages of simulation are the time required to design, write, debug and execute a sophisticated simulation, the effort necessary to validate the simulation, and the analysis required of the statistical simulation results. (Some of this effort can be reduced, however, if a simulation tool is used, where much of the programming and debugging effort will have been carried out by the tool authors.)

The statistical nature of the simulation results causes several difficulties. One of these arises from the correlation between the results of successive simulation runs that makes it difficult to form confidence intervals for the estimated performance measures. In addition, the statistical nature of the results makes it difficult to use simulation models as the basis for the optimisation of system parameters. Finally, given the statistical nature of simulation, large amounts of computer time (and expense) are required to characterise accurately the estimated performance measures.

3.2 Generating random numbers

One of the foundations on which any simulation programme is based is the generation of a sequence of (pseudo) random numbers.

As we will discover below, there is no way, short of tapping into some 'natural' source of random numbers, of generating genuinely random numbers on a computer. We will content ourselves, therefore, with generating a deterministic sequence of numbers whose properties approach as closely as possible to a sequence of purely random numbers.

From such a sequence of numbers $\{Z_i\}$ on range 0 to M, we will be able to derive a sequence of numbers $\{Z_i' = Z_i/M\}$, that appear to be a sequence of independent identically-distributed samples from the uniform probability density distribution u(x) on the range (0,1) shown in Figure 3.1.

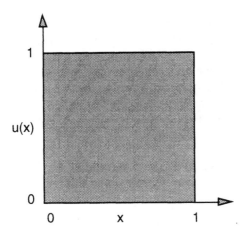

Figure 3.1 Uniform probability density distribution

Subsequently, we will see how we might transform this uniform distribution to produce the other probability distributions required for simulation purposes, such as the negative exponential and Poisson distributions.

3.2.1 Midsquares method

By way of introduction, and for historic perspective, let us consider the midsquares method. This approach was used long before the properties of such sequences were more clearly understood. The basic idea is a simple one; starting with a number (the seed), the number is squared, and then the central group of digits extracted to form the next random number, and so on.

The seed needs to be chosen very carefully, for example, if we start with a number, say **42**, as a seed, square it to give **1764**, so that the next number consists of the middle digits **76**. A series on these lines is then

$$42, 76, 77, 92, 46, 11, 12, 14, 19, 36, 29, 84, \ldots$$

Unfortunately if we start with a seed such as **23**, the series is

$$23, 52, 70, 90, 10, 10, 10, \ldots.$$

which is not very random! Even the first series degenerates in the next few terms. In practice, very large numbers would be used but similar effects are experienced.

3.2.2   *Linear congruential generators*

The next group of random-number generators will be the one we will focus on for the purpose of this chapter. The basic approach used is to start with an arbitrary number Z_0, which is then multiplied by an integer (a). The product is divided by another integer (M) leaving a remainder Z_1. The next term Z_2 is the remainder when aZ_1 is divided by M and so on. The generator is known as a multiplicative congruential generator. In some cases a further integer (b) may be added after the multiplication. The mechanism is then referred to as a mixed congruential generator. The recursion may be expressed as

$$Z_i \equiv (aZ_{i-1} + b) \ (\mathrm{mod}\ M)$$

$$(3.1)$$

The congruence sign means that Z_i is determined algebraically as

$$Z_i = aZ_{i-1} + b - Mk_i$$
$$\text{where } k_i = \mathrm{int}\,[\,(aZ_{i-1} + b)/M]$$

$$(3.2)$$

Starting with the initial seed, Z_0, we have

$$Z_1 = aZ_0 + b - Mk_1$$

$$Z_2 = a^2 Z_0 + ab - aMk_1 + b - Mk_2$$

$$Z_3 = a^3 Z_0 + b(1 + a + a^2) - M(k_3 + k_2 a + k_1 a^2)$$

etc.

Then

$$Z_n = a^n Z_0 + b(1 + a + \ldots + a^{n-1}) - M(k_n + k_{n-1}a + \ldots + k_1 a^{n-1})$$

$$\equiv a^n Z_0 + \frac{b(a^n - 1)}{(a-1)} \ (\text{mod } M)$$

$$(3.3)$$

It is obvious that for given values of constants, a, b and M, the seed Z_0 completely determines the sequence of numbers produced. In spite of this, if the values of the constants and the seed are chosen properly, it is possible to produce a pseudo-random sequence that is difficult to distinguish from a sequence of purely random numbers.

Period of numbers generated

Starting with the modulus M, the value of the constant a must be given a value such that $0 < a < M$. Then the pseudo-random variable Z_i can assume $P \leq M$ distinct values before the sequence is repeated. The quantity P is known as the period of the generator and this should be as large as possible. In Equation 3.3, if n = P, then $Z_n = Z_0$.

Then

$$Z_0 \equiv a^P Z_0 + \frac{b(a^P - 1)}{(a-1)} \ (\text{mod } M)$$

$$\Rightarrow (a^P - 1)\left[Z_0 + \frac{b}{(a-1)}\right] = Mk$$

$$\text{where } k = \sum_{i=1}^{P} k_i a^{P-i}$$

$$\Rightarrow (a^P - 1)\left[Z_0 + \frac{b}{(a-1)}\right] \equiv 0 \ (\text{mod } M)$$

$$(3.4)$$

The period of the generator is the minimum value of P that satisfies Equation 3.4.

The values of the constants a, b and M and the seed should be selected to maximise this minimal P.

There are several ways of selecting the constants and the seed. A useful approach is to specify $M = 2^\beta$, where β is the word size of the computer. Then for example, on a 16-bit machine, the largest integer that can be stored in a single word is $2^{16} - 1$. The use of a power of 2 for M also means that multiplication and division can be replaced by simple shift operations. This greatly speeds up calculations, an important consideration when a large number of random numbers have to be generated in a simulation. If the constant b is relatively prime to M and $a \equiv 1$ (mod 4) or, equivalently, $a = 1 + 4k$ for integral k, then $P = M$. In this case we refer to a full period generator.

For example, if we take the modulus $M = 16$ and constants $a = 9$ and $b = 11$, with seed $Z_0 = 3$ the next number is $Z_1 = ((9 \times 3) + 11) \bmod 16 = 6$ and so on. The series is given in Table 3.1 and it will be seen that it repeats after 16 terms.

Alternatively, still with $M = 2^\beta$ we can make $b = 0$, $a = \pm 3 + 8k$ or $a = 1 + 4k$ for integral k. The seed Z_0 must be chosen to have an odd value. This gives a period $P = 2^{\beta-2} = M/4$. In this case we refer to a maximal period generator.

Values of M other than 2^β may be considered. If M is the largest prime number in 2^β and $b = 0$, then it may be shown [3] that the period $P = M - 1$ if a is a primitive root of M. For a to be a primitive root of M, it must satisfy the equation

$$a^P = 1 + MK$$

where K is an integer and for any integer , $Q\ (< P)$

$$\frac{a^Q - 1}{M} \quad \text{is non-integral.}$$

Table 3.1 *Example of a full period random number generator*

i =	0	1	2	3	4	5	6	7	
Z_i =	3	6	1	4	15	2	13	0	

i =	8	9	10	11	12	13	14	15	16
Z_i =	11	14	9	12	7	10	5	8	3

This type of generator is known as a prime modulus multiplicative congruential generator. An advantage is that the seed is not restricted to an odd number as for other types of multiplicative generators.

It is not easy to find the primitive roots of the prime moduli. However, we may note that if a is a primitive root, then a^K is also a primitive root provided K is relatively prime to P. The factorisation of P is useful in complying with the restriction on K. For example, the prime modulus $M = 2^{31} - 1$ is of particular interest. The factorisation of its period is

$$P = 2^{31} - 2 = 2 \times 3^2 \times 7 \times 11 \times 31 \times 151 \times 331$$

This modulus is used in some commercially available simulation programs [3].

3.2.3 Additive congruential generators

Additive congruential generators are intended to allow the period of the generator to be extended beyond 2^β. The procedure is to first generate a sequence of N random numbers, using one of the methods described above, and then to use this initial sequence as seeds for the full set of random numbers. The simplest application of the method is to generate a new number in the sequence by adding the last number generated to the Nth previous number, that is

$$Z_i \equiv (Z_{i-1} + Z_{i-N}) \bmod M$$

$$(3.5)$$

If appropriate weighting factors, c_j are chosen, it is possible to achieve periods as long as M^{N-1} for the general case, that is

$$Z_i \equiv \sum_{j=1}^{N} c_j Z_{i-j} (\bmod M)$$

$$(3.6)$$

For example, if we take the first seven numbers given in Table 3.1 (3, 6, 1, 4, 15, 2, 13) and apply the procedure with $N = 7$, we get the series shown in Table 3.2.

Table 3.2 Example of n additive congruential generator

3	6	1	4	15	2	13
0	6	7	11	10	12	9
9	15	6	1	11	7	• 0
9	8	14	15	10	1	1
10	2	0	15	9	10	11

and so on

An extension of this technique is to generate a stream of random bits rather than integers [4]. The integers are then derived from groups of bits. The generator is then known as a linear recurrence generator modulo 2. The constants, c_j are chosen so that the recurrence relationship takes the form

$$B_i \equiv B_{i-Q} + B_{i-N}(\text{mod } 2) \equiv B_{i-Q} \oplus B_{i-N}$$

$$(3.7)$$

where B_i is used to represent bits as distinct from integers and $\oplus$ represents the 'exclusive OR' function. Hardware to implement the above function is relatively simple and generation of random numbers may be carried out very quickly.

3.2.4 Computational considerations

In many simulation programs, a substantial part of the time taken to run the programme is occupied in the generation of random numbers. If we consider Equation 3.1 it will be seen that the division by M to obtain the largest possible integer (k_i) and then the multiplication of k_i by M are the steps likely to take the longest time to compute. The time is greatly reduced if the divisor or multiplicand is a power of 2 so that the division or multiplication may be achieved by simple shifting operations.

In order to extend the advantage of simple shifting operations instead of multiplication and division to prime-modulus multiplicative congruential generators, we proceed as follows.

Consider the case where M is the largest prime number in 2^β. As an example take the case of $M = 2^{31} - 1$ mentioned above.

Let

$$k = int[a\ddot{Z}_{i-1}/2^{31}]$$

(3.8)

denote the largest positive integer in $a Z_{i-1}/2^{31}$ so that

$$\ddot{Z}_i = a\ddot{Z}_{i-1} - k\,2^{31}$$

(3.9)

If $\ddot{Z}_i + k < M$ then

$$Z_i = \ddot{Z}_i + k \equiv a\ddot{Z}_{i-1} \pmod{M}$$

otherwise

$$Z_i = \ddot{Z}_i + k - M \equiv a\ddot{Z}_{i-1} \pmod{M}$$

(3.10)

3.2.5 An example generator

Our difficulty now, is how to transform the above equations into C++ code for use on both 16- and 32-bit machines. The main problem is that while a standard representation exists for 32-bit variables in C++, i.e. the 'unsigned long' integer, there is no such quantity for 64-bit. Consequently, when we come to multiply the 31-bit variable representing Z_{i-1} with the possibly 31-bit quantity representing A, we will have no way of representing the product other than by the use of multiple 'longs'.

We must therefore transform the above equations into a suitable form for implementation using longs.

Let

$$Z_{i-1} = [Z_{i-1}]_{hi}\,2^{16} + [Z_{i-1}]_{lo}$$

and

$$a = [a]_{hi}\,2^{16} + [a]_{lo}$$

Then

$$aZ_{i-1} = \{[a]_{hi}\,2^{16} + [a]_{lo}\}\{[Z_{i-1}]_{hi}\,2^{16} + [Z_{i-1}]_{lo}\}$$

$$= [aZ_{i-1}]_{hi}\,2^{32} + [aZ_{i-1}]_{mi}\,2^{16} + [aZ_{i-1}]_{lo}$$

(3.11)

Hence

$$k = int [a Z_{i-1}/2^{31}]$$

$$= int [[aZ_{i-1}]_{hi} 2 + [aZ_{i-1}]_{mi} 2^{-15} + [aZ_{i-1}]_{lo} 2^{-31}]$$

(3.12)

and

$$\overset{..}{Z_i} = a\overset{..}{Z}_{i-1} - k2^{31}$$

(3.13)

If $Z_i + k < M$ then

$$Z_i = \overset{..}{Z_i} + k \equiv a\overset{..}{Z}_{i-1} \ (mod \ M)$$

otherwise

$$Z_i = \overset{..}{Z_i} + k - M \equiv a\overset{..}{Z}_{i-1} \ (mod \ M)$$

(3.14)

The equations for k and Z_i are implemented using shifts, and they can perhaps best be understood diagrammatically in Figure 3.2.

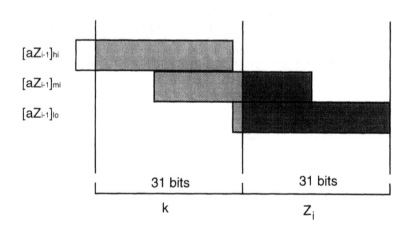

Figure 3.2 Shifts in expressions for computation

Thus k is equal to the lower 30 bits of $[aZ_{i-1}]_{hi}$ left shifted by 1, plus the upper 17 bits of $[aZ_{i-1}]_{mi}$ right shifted by 15, plus the upper bit of $[aZ_{i-1}]_{lo}$ right shifted by

31 places. Similarly, Z_i is equal to the lower 15 bits of $[aZ_{i-1}]_{mi}$ shifted left by 16, plus the lower 31 bits of $[aZ_{i-1}]_{lo}$.

This may be coded using the C++ language.

```
/************************************************************

    urand5()

    A linear congruential random-number generator,
    of prime modulus multiplicative type, with M = 2**31-1
    and A = 397204094. The numbers produced are in the
    range 0+ to 1.

************************************************************/

#define RAND_M   0x7fffffffL
#define RAND_AHI 0x17acL
#define RAND_ALO 0xda7eL

static double urand5(unsigned long* pz) {
    unsigned long zhi, zlo, zahi, zami, zalo, k;

    zhi = (*pz & ~0xffffL) >> 16;
    zlo = *pz & 0xffffL;

    zahi = zhi * RAND_AHI;
    zami = zhi * RAND_ALO + zlo * RAND_AHI;
    zalo = zlo * RAND_ALO;
    k   = (zahi << 1) + ((zami & ~0x7fffL) >> 15) +
              ((zalo & ~0x7fffffffL) >> 31);

    *pz = ((zami & 0x7fffL) << 16) +
              (zalo & 0x7fffffffL) + k;

    if (*pz >= RAND_M)
          *pz -= RAND_M;

    return((double) *pz/(double) RAND_M);
}
```

3.2.6   *Empirical testing of generators*

A simulation run is effectively a sample taken from a larger population. It is important to ensure that the conditions for generating the sample are sufficiently close to the system that is being modelled by the sample. Hence, we must have confidence in our pseudo-random number generator. A large number of empirical tests exist, including several tests of the empirical DF, the run-up and -down test, the chi-squared test, the serial test and several tests of correlation. We will briefly consider two of these.

Chi-squared test

In sampling theory, a constructed random variable, chi-squared (χ^2) is used to determine whether a given sample is likely to have been derived from a given larger population whose statistics are known or are assumed. If the value of χ^2 is low enough, we can say with probability $1 - F(\chi^2)$ that the sample is consistent with having been drawn from the total population. If we continued sampling, the sample distribution would tend towards the population distribution and χ^2 would tend to zero.

The random variable χ^2 is defined in terms of the frequency plots for the sample of the assumed distribution. If the number of experimental outcomes in sub-range n is O_n and the number of occurrences predicted by the assumed distribution for the same sub-range is E_n then

$$\chi^2 = \sum_{n=1}^{N} \frac{(O_n - E_n)^2}{E_n}$$

where N is the total number of sub-ranges.

(3.15)

The distribution function for the χ^2 random variable is available in statistical tables. Some common values are given in Table 3.3.

Table 3.3 Values of the chi-squared distribution

nF	0.01	0.05	0.10	0.90	0.95	0.99
degrees of freedom						
7	1.24	2.17	2.83	12.02	14.07	18.48
8	1.65	2.73	3.49	13.36	15.51	20.09
9	2.09	3.33	4.17	14.68	16.92	21.67
10	2.56	3.94	4.87	15.99	18.31	23.21
20	8.26	10.85	12.44	28.41	31.41	37.57
30	14.95	18.49	20.60	40.26	43.77	50.89

The degrees of freedom represent the number of different ways the sample and assumed distributions may vary from one another. If there are a large number of ways in which the distributions may differ, a larger value of χ^2 may be expected. The values on the top of the table are the probabilities of $\chi^2 \leq x$ from the distribution function. For the given number of degrees of freedom, the values given in the table are the probabilities that the actual value of χ^2 would fall below the value x is the value in the column heading where x was found. This represents the confidence level at which we can reject the hypothesis that the sample distribution differed from the assumed distribution. It should be noted that the sample sequence size L should be large (10 000 say) to ensure that the test is fairly applied.

As an example in the use of the chi-squared test, a sample of 100 000 discrete variables in the range 0 to 9 was derived using the 'rand' function in a 'C' compiler and a seed derived from the number of seconds elapsed since midnight. (This is a very unsafe way of selecting the seed). The frequencies obtained in three successive runs were:

0	1	2	3	4	5	6	7	8	9
9960	10058	10041	10106	10013	10033	9945	10034	10003	9807
10050	10000	10210	9946	10036	10008	9799	9970	9999	9982
9946	10064	9907	9900	10032	9970	10047	9912	10087	10135

The values of χ^2 calculated from Equation 3.15 are found to be 6.06, 9.52 and 6.33 respectively. Although the variable can take any one of ten values, there are only nine degrees of freedom since if nine of the frequencies are set, the final one is fixed. From the table, only if the value of χ^2 is 4.17 or less can we be $(1 - 0.1) =$

90% confident that the sample is drawn from a population with a uniform distribution. On the other hand, the value of χ^2 would have to be 14.68 or greater for 90% confidence that the sample was not drawn from a uniformly distributed population. These results show that sequences of random numbers derived in this way should be regarded with some caution.

So how do we apply all this to the testing of random-number generators? Well, the two distributions being compared are the uniform distribution and the unknown distribution of our random-number generator. The degrees of freedom in the general case is the number of sub-ranges N, minus the number of true differences in the density parameters R. In the case of the uniform distribution, R = 1. With a sample sequence of L random numbers $\{Z_i' = Z_i/M\}$, the expected number in each sub-range $E_n = L/N$, and

$$\chi^2 = \frac{N}{L} \sum_{n=1}^{N} (O_n - L/N)^2$$

(3.16)

A value X greater than or equal to χ^2 is then sought in the table on row N − 1, and the hypothesis that the sequence has a uniform distribution can be accepted with confidence level 1 − F, where F is the probability at the head of the column in which X was found. One point that should be stressed is that, if the sequence of numbers involved in the chi-squared test is not independent then, in principle, the test is invalid. In practice, if the covariance is small or positive, then the test will perform satisfactorily; however, negative covariance of any size will invalidate the test. Thus, as with all empirical testing of random-number generators, it is important to use a series of tests gradually to build confidence in the performance of the generator, rather than accepting it on the basis of one or even a few results.

Checking for independence
A series of pseudo-random numbers should not only be consistent with a uniform distribution but also each number should be independent of its predecessors. Since each number is derived from a previous number or numbers this cannot be absolutely true, but it is possible to apply tests to check whether the series appears to consist of independent numbers. This is done by considering groups of numbers rather than individual ones. For example, if we consider the sequence of numbers given in Table 3.2, the two-dimensional vectors formed would be (3,6), (1,4), (15,2), (13,0) . . . etc. If the numbers were independent, we would expect the vectors to be uniformly distributed within the square (0,0), (0,15), (15,0), (15,15) and a chi-squared test can be applied to verify this.

3.3   Other probability distributions

In some simulation exercises we may need to generate sequences of numbers that represent random variables, but have probability distributions other than the uniform distribution discussed above. For example, it may be necessary to simulate the effects of human response times or variations in component values due to manufacturing tolerances. In these cases, the parameters are likely to have a distribution about a mean that is approximately normal. There are three main approaches to the generation of suitable sequences of numbers, the Inverse Transform Method, the Composition Method, and the Accept-Reject Method.

3.3.1   Inverse transform

This approach is based on the use of a function which is the inverse of the required distribution function. Starting with a distribution function $F(y)$ for the random variable Y, a new random variable may be formed, $Z = F(Y)$. The distribution function for the new random variable,

$$G(z) \; = \; P[Z \leq z] \; = \; P[\, F(Y) \leq z]$$

$$(3.17)$$

Assuming that the inverse function F^{-1} exists, the events $F(Y) \leq z$ and $Y \leq F^{-1}(z)$ are the same. Hence

$$G(z) \; = \; P[\, Y \leq F^{-1}(z)]$$

$$= \; F(F^{-1}(z)) = z \qquad\qquad 0 \leq z \leq 1$$

$$(3.18)$$

$$\text{since} \quad P[Y \leq y] \; = F(y)$$

Then since Y is distributed according to the required distribution function $F(y)$ the variable Z will be uniformly distributed in the range 0 to 1. It follows then that random numbers with the distribution function $F(y)$ may be produced by first generating random numbers with a uniform distribution, and then applying the inverse transformation of $F(Y)$, that is, $F^{-1}(z)$ to those numbers. Unfortunately, an inverse transform does not always exist, although in some cases it is possible to perform some preliminary manipulation to give an expression that can be transformed. If this is not possible, then other methods must be used.

An example of the technique is the production of a negative exponential distribution.

Negative exponentially distributed random numbers
In this case, the distribution function is monotonically increasing, so the inverse
exists.

$$F(y) = 1 - e^{-\lambda y}$$
$$\Rightarrow z = 1 - e^{-\lambda y}$$
$$\Rightarrow e^{-\lambda y} = 1 - z$$
$$\Rightarrow -\lambda y = \ln(1 - z)$$
$$\Rightarrow y = -\frac{1}{\lambda} \ln(1 - z)$$

(3.19)

If a sequence of uniformly distributed random numbers is substituted in turn into
Equation 3.19, the result will be a sequence of exponentially distributed random
numbers. A simplification is possible, since if z is uniformly distributed, $(1 - z)$
will also have a uniform distribution.

$$y = -\frac{1}{\lambda} \ln(z)$$

(3.20)

An implementation of this method in C++, using the function 'urand5 ()' is then:

```
/ * * * * * * * * * * * * * * * * * * * * * * * * * * * * * * * * * * * * * * * * * * * * * * * * * * * * * *

    expgen()

    Generate negative exponential random numbers, mean
    1/lambda

* * * * * * * * * * * * * * * * * * * * * * * * * * * * * * * * * * * * * * * * * * * * * * * * * * * * * */

double expgen(double lambda) {
    static double urand5(unsigned long* pz);

    assert(lambda > 0.0);

    double v = -log(urand5(&seed));
    return(v/lambda);
}
```

The generation of a number having a normal distribution is more difficult using the inverse method, since it is impossible to obtain a closed form for the expression. However, with a little manipulation, a suitable expression may be found.

Normally distributed random numbers
However, it may be shown [13] that if Z_1 and Z_2 are two independent random variables in the range 0 to 1 and

$$V_1 = 2Z_1 - 1$$

$$V_2 = 2Z_2 - 1$$

(3.21)

and also

$$R^2 = V_1{}^2 + V_2{}^2 \qquad R < 1$$

then

$$N_1 = V_1 \frac{\sqrt{-2 \ln R^2}}{R}$$

$$N_2 = V_2 \frac{\sqrt{-2 \ln R^2}}{R}$$

(3.22)

where N_1 and N_2 are two random numbers from a normal distribution which has mean = 0 and variance = 1. Each calculation yields two normally distributed random numbers provided $R < 1$. If R is greater than 1, the result must be rejected and the calculation re-started.

An implementation in C++ is then:

```
/*******************************************************

   normgen()

   Generate normally distributed random numbers, mean mu,
   standard deviation rho

*******************************************************/

double normgen(double mu, double rho) {
```

```
static double urand5(unsigned long* pz);
static bool   oddcall = true;
static double stored, returned;

assert(rho > 0.0);

if (oddcall) {                    // no normal number stored;
  //use separate random number streams to ensure
  //independence
  double v1  = 2.0 * (urand5(&seed1)) - 1.0;
  double v2  = 2.0 * (urand5(&seed2)) - 1.0;
  double rsq = (v1 * v1) + (v2 * v2);

  while (rsq > 1.0) {
    v1  = 2.0 * (urand5(&seed1)) - 1.0;
    v2  = 2.0 * (urand5(&seed2)) - 1.0;
    rsq = (v1 * v1) + (v2 * v2);
  }

  double k  = sqrt((-2.0 * log(rsq))/rsq);
  double n1 = v1 * k;
  double n2 = v2 * k;

  returned = n1;
  stored   = n2;
  oddcall  = false;
}
else {                    // even call; use stored normal number
  returned = stored;
  oddcall  = true;
}

return(mu + rho * returned);
}
```

3.3.2   Composition

Some probability distributions can be defined in terms of other distributions. Then, if we can generate sequences having the underlying distributions, we can then construct the required 'composite' distribution. This technique may be used to generate sequences of numbers having a Poisson distribution.

Poisson distribution

In Chapter 2 it was shown that for a random arrival (Poisson) process the number of arrivals within a fixed time interval T has a Poisson distribution and that the inter-arrival distribution is negative exponential. The two distributions may be combined by counting the number of exponentially distributed inter-arrival times to add up to the time interval T. The result is a sequence of numbers having a Poisson distribution

Such a routine, implemented in C++ is given below:

```
/*******************************************************

    poisson1()

    Generate a Poisson distribution for a Poisson process
    of arrival rate lambda and time interval t i.e. mean
    lambda*t.  The method used is composition from the
    negative exponential distribution.

********************************************************/
long poisson1(double lambda, double t) {
    static double urand5(unsigned long* pz);
    assert(lambda > 0.0);
    assert(t > 0.0);

    long r = 0;       //cumulative no. of arrivals in
                      //interval t
    double time = -log(urand5(&seed))/lambda;
                      //sum of inter-arrival times of r+1
                      //arrivals
    while (time <= t) {
       r++;
       time += -log(urand5(&seed))/lambda;
    }
    return(r);
}
```

The above routine will correctly generate the required Poisson distribution, but by recognising that summing a series of logs is equivalent to multiplying their contents, and by rearranging some of the expressions, a far more efficient routine can be implemented:

```
/ * * * * * * * * * * * * * * * * * * * * * * * * * * * * * * * * * * * * * * * * * * * * * * * * * * * * *

   poisson2()

   Generate a Poisson distribution for a Poisson process
   of arrival rate lambda and time interval t i.e. mean
   lambda*t. The method used is composition from the
   negative exponential distribution, and this version
   takes advantage of certain properties of logs to
   reduce execution time compared to poisson1().

* * * * * * * * * * * * * * * * * * * * * * * * * * * * * * * * * * * * * * * * * * * * * * * * * * * * * /

long poisson2(double lambda, double t) {
    static double urand5(unsigned long* pz);
    assert(lambda > 0.0);
    assert(t > 0.0);

    long r = 0;        //cumulative no. of arrivals in
                       //interval t
    double exp_lambda_t   = exp(-lambda*t);
                       //transformed t
    double exp_lambda_time = urand5(&seed);
                       //transformed sum of inter-arrival
                       //times of r+1 arrivals

    while (exp_lambda_time >= exp_lambda_t) {
      r++;
      exp_lambda_time *= urand5(&seed);
    }
    return(r);
}
```

3.3.3 Accept-reject

The accept-reject approach can often be used if the other methods are not available. Consider a variable X having a density f(t), which cannot be generated by other methods. Then generate a variable Y having density g(t) so that for all t, c g(t) $\geq$ f(t), where c is a constant. It is usually convenient to make g(t) a uniform density function such that c is equal to the maximum value of density f(t). A third variable U may be generated with uniform density but independent of Y.

If $\quad\quad\quad U\ g(Y) \le\ f(Y)\ /c$

then accept the value of Y as the next value of X, otherwise reject and start again. For example, for f(t) bounded by the shaded area in the graph in Figure 3.3, with c=1 and g(t)=u(t), the uniform density distribution, accept Y as the next generated value of X if U ≤ f(Y).

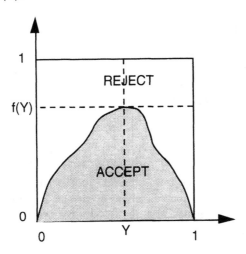

Figure 3.3   Acceptance or rejection of value of Y

Clearly, in general, a careful choice of c and g(t) must be made to avoid excessive computation.

3.4   Programming simulations

When a simulation programme is run, operations occur which result in a change of state in the model of the system. Examples of such operations are the arrival of a packet, the completion of a telephone call or the failure of a component and they are known as events. The whole purpose of the simulation programme is to follow a sequence of events and so estimate the performance statistics of the system.

Although the simulation models the events of the real system, it does so on a completely different time scale. For example, it may be required to model 24 hours operation of a system. This is known as the simulation time. The run time or execution time, that is the amount of time in executing the simulation programme, may be only a few minutes. In general, the run time is very much shorter than the simulation time although in some cases, such as modelling a nuclear bomb explosion, the run time is very much longer! Note also that whereas time flows evenly during the operation of the real system, simulated time is very erratic.

There are two basic approaches to writing a simulation programme. The first, time-based simulation is the simpler but it is relatively inefficient. The second, event-based simulation (or discrete-event simulation) is generally more efficient and more accurate.

3.4.1   Time-based simulation

The inefficiency of time-based simulation may be caused by having to deal with many clock intervals in which nothing occurs or by having a high probability of an event at clock tick [6]. The former case is particularly noticeable in, for example, reliability simulations in which a component failure is a very rare event. However, as computer power increases, the inefficiency of a time-based system becomes less important.

A high-level flowchart for a time-based simulation is shown in Figure 3.4. For each execution of the main control loop, the simulation clock is advanced by one unit of time, δt. At each such tick, a random number is normally generated to determine whether one or more events have occurred. The events are processed if necessary and there is the underlying assumption that the order of events within a time interval is unimportant.

3.4.2   Event-based simulation

In event-based simulation, the execution of the control loop does not occur at regular intervals of (simulated) time, but moves on immediately to the next event. This makes for greater efficiency, since no time is wasted in dealing with time intervals in which there are no events. After each event the simulation clock is then advanced the amount of time that has elapsed since the previous event. Only one event can be processed at any one pass, which simplifies the high-level control structure. A flow-chart of an event-based simulation is shown in Figure 3.5

A problem in programming event-based simulations is the determination of which event is next. The complication arises because the processing of one event could give rise to several other events. The information about those events is stored in an event queue to enable them to be scheduled. Although events are normally stored in the event queue in time order, this may not be necessary [6]. Furthermore, the processing of an event may generate other events that interleave in time with events already in the queue. In an extreme case, newly generated events may lead to the removal of events already in the queue. An example of this would be the failure of a switch matrix in a telephone exchange, which would imply the elimination of all the calls associated with that switch.

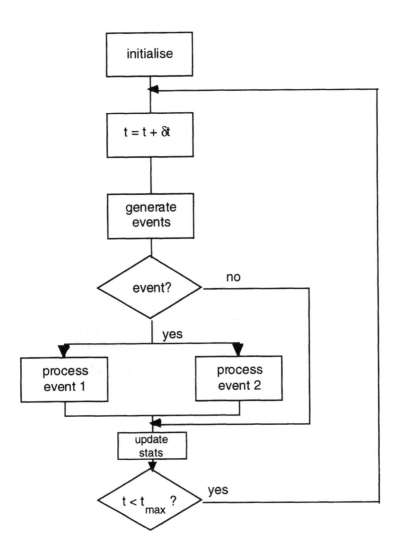

Figure 3.4   Flow chart for time-based simulation

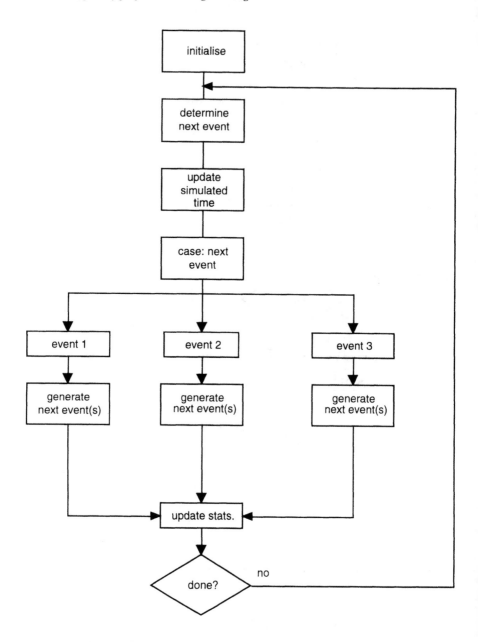

Figure 3.5   Flow chart for event-based simulation

3.4.3   Statistics of simulation

The main objective of a simulation programme is to learn how the real system would operate under the given conditions. It would, of course, be possible to print out all the simulated events and the state of the model after each simulated event. This is a useful technique when debugging or verifying a simulation programme, but it would require considerable processing of the output to obtain useful results. It is far preferable to generate the information necessary to assess the performance of the real system as the simulation proceeds. The statistics gathered take the form of one or more frequency plots, corresponding to the probability of a theoretical analysis, and/or summary statistics.

It is important to decide at the outset what summary statistics are necessary. If this is not done, it will be necessary to revise the programme and then to run the whole simulation again to obtain the necessary information. For example, in a packet network it may be desirable to assess not only the fraction of packets lost through buffer overflow but also how the statistic would be affected if all the buffer stores were increased in size by given amounts. A further problem may be the numerical accuracy. For long simulations, the accumulation of statistics may lead to problems of arithmetic overflow. The limited precision of some machines can also cause trouble. For this reason, it is often advisable to produce one or more frequency plots to provide a check on the summary statistics even if they are not required directly. Unfortunately, if the frequency plots are sufficiently detailed, they require large amounts of memory.

3.4.4   Analysing and verifying simulation results

Simulation programme verification

Having programmed your simulation programme, based on a good random-number generator, including efficient functions for the production of the appropriate probability distributions, and with your best effort at untangling the complexities of the event sequencing, how do you actually know it is correct? Well, the answer is you don't! But two important techniques can help build confidence in the programme. The first is the use of assertions in the programme code; in developing the simulation programme as a model of the real system, all the otherwise unconscious assumptions made should be included as assertions. For example, when simulating an M/G/1 queue, say, service cannot begin unless there is a packet waiting to be served. It might well be possible to write a simulation programme for this system such that packet service began even though there was not a packet waiting to be served. This could happen through the queue being written in such a way as to allow a packet to be removed from it even when one was not present; the number of packets in the queue would then be negative, and any subsequent arrivals at queue would have been served before they arrived (until the number on the queue went positive again)! This sort of difficulty could be avoided by asserting that a packet must be present before service can begin, or that the queue size must not be negative. For efficiency reasons, such checks can be removed, by conditional compilation, before making extensive use of the simulation programme. The second technique is to run the simulator, perhaps under restricted or boundary conditions, on a problem that can be solved analytically (or for which simulation results are given in the literature). Comparison of the simulation results statistically, over a number of runs, with the analytical ones can help to establish confidence in the correctness of the simulation programme before using it to obtain new results.

Start-up conditions

In some cases of simulation, it may be required to assess the performance when the system is starting from 'cold'. An example of this would be the behaviour of a telephone exchange which is restored to operation after a complete breakdown when all calls were lost. However, in most cases we are concerned with the operation when the system has been running normally for some time. In the case of the telephone exchange, it is usual to consider the performance during the busy hour when calls have already built up to a high level.

If the simulation is run for a very long time, it may be possible to assume that the start-up behaviour has negligible effect on the statistics obtained. This is expensive in computer time and there must inevitably be an element of uncertainty in ignoring the effects of start-up. A preferable approach is to run the system for a time to ensure steady-state conditions and only start accumulating statistics after this time has elapsed. Figure 3.6 shows a typical plot of the number of calls in progress as a telephone exchange is started from cold. Then for further simulation runs of the system, all events before t_s are ignored.

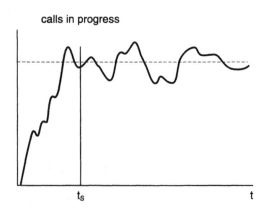

Figure 3.6   Start-up conditions for telephone exchange traffic

An alternative method is to start the simulation by setting the parameters to the average or near average values that would be expected after the system had been running for some time, or to start the system at a regeneration point for the stochastic process (see below). Although this latter technique, known as regenerative simulation, may seem very attractive, it is difficult to apply as we shall see later.

Confidence intervals

Because of the manner of generating random events using a pseudo-random number generator, as described earlier in this chapter, it follows that repeated runs using the same constants and seed will produce identical events and statistics. Although this is an advantage when debugging the programme, it does not help us if we wish to obtain additional data.

In general, no one sample path for a stochastic process will correspond to the mean behaviour of the system. How then do we obtain the mean behaviour for the underlying stochastic process from a simulation run which only represents a particular sample path? The answer is not to use one simulation run, but several, and thereby to obtain an estimate of the error in the experimental mean.

Consider a simulation of a packet system with a limited queuing buffer. Each run involved 10 000 packets and the numbers of packets lost were:

run	packets lost
1	88
2	96
3	107
4	94
5	90

No two runs resulted in the same number of packets lost, so what value should we use for the mean delay? Clearly, we could take the mean of the results, which is 95, but given the amount of variation, what is the potential error in such an estimate?

As in any collection of samples from a large population, it is useful to know how representative the samples are likely to be. Clearly our best estimate of the mean of the population is the average of the sample means. It is assumed that the collection of n results from the simulations x_k are normally distributed about this mean with a sample variance (s^2) calculated from the simulation runs. Then it is possible to estimate the interval $\pm \varepsilon$ within the population mean would fall with probability p from the Student-t distribution:

$$\varepsilon = t_{n-1, p} \sqrt{\frac{s^2}{n}}$$

(3.23)

The Student-t distribution is given in statistical tables. A small subset is given in Table 3.4.

In the packet example given above, suppose it is required to calculate the 90% confidence interval. The experimental mean of the 5 runs is seen to be 95 and the variance 55. Now in Table 3.4, with $5 - 1 = 4$ degrees of freedom and probability $p = 0.90$, the value of $t_{4,\ 0.90}$ is 2.13. Then with 90% confidence, the mean number of packets lost in 10 000 is 95 ± 7.1. If we require 98% confidence, we can only say the mean number of packets lost is 95 ± 12.

Table 3.4 Some values of the student-t distribution

	Confidence Probability (p)		
n	0.90	0.96	0.98
4	2.13	3.00	3.75
5	2.02	2.76	3.37
9	1.83	2.40	2.82
10	1.81	2.36	2.76
100	1.66	2.08	2.36

It is not of course necessary to run the simulation a number of times. A long simulation run may be broken into n shorter batches and separate statistics recorded for each batch. This approach reduces some of the problems of start-up transients since the system conditions at the end of one batch form the conditions for the start of the next batch. This involves the assumption that the batches are independent. The further assumption in forming confidence intervals is that the means of a number of simulation batches are normally distributed, and this is not always justified, particularly if the number of batches is small. If the simulation is run for a long time and the statistics for many batches are measured, the Student-t approach is invaluable for estimating how good our results are likely to be. The problem is knowing how long is a sufficiently long time. This of course is the weakness of all simulation. We can never be sure of the results and in this case, we can only say that the confidence intervals are probably right.

Regenerative simulation
The stochastic process modelled in a simulation will normally have points at which the process continues with the same distribution as it started with. These points are called regeneration or renewal points. If the simulation programme can be coded to make the batches correspond with the intervals between regeneration points, then the

batch samples are independent regardless of their length. It follows then that if the number of intervals is large, the results derived from the batches will tend towards a normal distribution. The problem is to find the regeneration points. Although the use of regeneration points removes the need to make the assumption that the batches are independent, there remains the assumption that the statistics measured are normally distributed. In the estimation of confidence intervals, the latter assumption is often the more crucial.

3.5 An example simulation programme

In order to illustrate some of the concepts presented in this chapter, let us now consider a small example programme. This is an event-based simulation coded in C++ [9,10], an object-oriented extension [11,12] of the perhaps more familiar 'C' programming language. For the sake of simplicity, no attempt has been made to cope directly with start-up transients. Also, in order to keep the size of the example down, and to avoid distracting the reader with unfamiliar constructs, the code is not perhaps as 'object-oriented' as it could be; one of the features and, some would argue, one of the faults of C++, is that it is a superset of 'C', and will thus allow the compilation of most ANSI 'C' programs as valid C++. Consequently, it is quite possible to code an example, as has been done here, in a less than purely object-oriented fashion. No discussion of the C++ programming language will be given here, although readers familiar with C++ should not find the code too difficult to follow.

3.5.1 The subscriber's private meter example

The example programme simulates a subscribers private meter (SPM) source. Historically, an individual telephone subscriber who wished to be advised of the number of units charged as a telephone call is proceeding, could arrange to have a subscriber's private meter installed by the administration. This meter would record the units charged to the subscriber by counting metering pulses sent by the telephone exchange, over the voice circuit, whenever the exchange calculated that a further unit had been spent. The metering pulses, which in the U.K. were 0.25 seconds in duration, would be supplied by an SPM source or sources on the exchange. Such an SPM source, however, would have only a finite capacity, and thus would only be able to generate a certain number of meter pulses simultaneously before the electrical characteristics of the pulses deteriorates to the point that the individual SPMs can no longer detect them on arrival. A typical configuration might have a single SPM source, capable of generating up to 64 simultaneous pulses, to serve a concentrator with up to 2048 subscribers.

What we wish to determine is, given a certain level of pulse requests, generated by the exchange processor, and sent to the SPM source, the probability of the SPM source being asked to generate more than **64** pulses at the same time (see Figure 3.7). To simplify the model, we will assume that the SPM source will block any further requests received when it is already producing **64** meter pulses at the same time, and that the pulse requests arrive at random intervals. (This is true in the limit, but on any one call, the pulse requests will arrive deterministically, i.e. separated by fixed intervals). The situation is thus equivalent to a random arrival process of mean rate λ applied to a loss system of 64 servers with deterministic service time $1/\mu$ of 0.25 seconds, and thus has a blocking probability given by the Erlang B formula, i.e. B(64, 0.25λ).

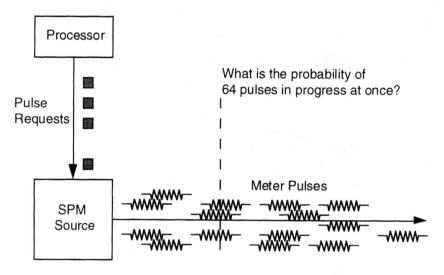

Figure 3.7   Generation of meter pulses for subscriber's private meter

The SPM programme

The programme consists of four files: `spm.h`, `spm.C`, `source.C` and `event_q.C`. The first of these, `spm.h`, is a header file containing the class declarations for the five classes used in the programme: `source`, `event_q`, `event`, `request_arr` and `pulse_end`. The other three files are C++ source files, containing respectively, the main function, the member functions for class `source`, and the member functions for class `event_q`. The member functions for the `event`, `request_arr` and `pulse_end` classes are given inline in the header file `spm.h`. Let us now look at each of these files in turn.

The SPM.h *header file*

After including the necessary standard header files, a new enumerated type, bool, is defined, the expgen function defined in section 3.3 is declared, followed by three global variables: the initial and current pseudo-random number generator seeds (init_seed and seed), and the event queue, evq. Five classes are then declared.

The source class has a constructor, a member function for the main simulation loop (simul), and one to print out the final results of a run (print_out). In addition, it has five private member functions: busy, mainly used in assertions, and four functions corresponding to the four main events, i.e. req_blocked, pul_start, req_arrival and pul_end. Of the six data members, three define important aspects of an SPM source (arrival_rate, pulse_length and capacity), whereas the other three are for the accumulation of statistics (num_pulses, num_requests and num_req_blocked).

The event_q class represents an event queue as a circularly linked list of event objects (last points to the last event in the queue). There are size events in the queue, and the current simulation time is stored in q_time. Five public member functions are provided: a constructor and destructor, time to retrieve the q_time, join to add a new event to the time-ordered queue, and action to remove and action the first event on the queue.

The event class itself is an abstract class, that simply records the event time (e_time) and a pointer to the next event in the queue. Two concrete derived classes of event are then declared: request_arr and pulse_end. Each of these retains a pointer to the relevant source object (spm), as well as redefining the virtual member function action.

```
/*
   spm.h
*/

#ifndef _spm_h
#define _spm_h 1

#include <iostream.h>
#include <strstream.h>
#include <assert.h>
#include <math.h>

//FUNCTION DECLARATIONS
extern double expgen(double lambda);
```

```
//GLOBAL VARIABLE DECLARATIONS
class   event_q;
extern event_q evq;
extern unsigned long seed, init_seed;

//CLASS DECLARATIONS
class request_arr;
class pulse_end;

class source {   //SOURCE
  friend request_arr;
  friend pulse_end;
public:
          source(double ar, double pl, int c);
  void    simul(long max_requests);
  void    print_out(void);
private:
  bool    busy(void) { return (num_pulses < capacity) ?
                          false : true; }
  void    req_blocked(void);
  void    pul_start(void);
  void    req_arrival(void);
  void    pul_end(void);

  double arrival_rate;
  double pulse_length;
  int    capacity;
  int    num_pulses;      //current number of pulses
  long   num_requests;    //total pulse requests
  long   num_req_blocked; //total blocked requests
};

class event;

class event_q {   //EVENT QUEUE
public:
          event_q(void);
          ~event_q(void);
  double time(void) { return q_time; }
  void   join(event* e);
  void   action(void);
private:
```

```
  double q_time; //simulation time clock
  event* last;    //event queue
  long    size;
};

class event {   //EVENT
  friend class event_q;
public:
            event(double et)
                { assert(et >= 0.0); e_time=et; }
  virtual ~event(void) {}
  virtual void action(void) {} //action to perform
                              //when event occurs
protected:
  double  e_time; //event time
  event*  next;
};

//REQUEST ARRIVAL EVENT
class request_arr : public event {
public:
            request_arr(double et, source* s) : event(et)
                { assert(s); spm = s; }
  void    action(void) { spm->req_arrival(); }
private:
  source* spm;
};

class pulse_end : public event {   //PULSE END EVENT
public:
            pulse_end(double et, source* s) : event(et)
                { assert(s); spm = s; }
  void    action(void) { spm->pul_end(); }
private:
  source* spm;
};

#endif _spm_h
```

The SPM.C source file
This file simply contains the main function of the programme. First the five required command line arguments are processed: the arrival rate, the pulse length,

the source capacity, the maximum number of pulse requests, and the seed for the pseudo-random number generator. Then a source object, spm, is created, the simulation loop is performed the requested number of times, and finally, the results are printed out.

```
/*
  spm.C

  Simulation to determine the grade of service
  of an SPM source.
*/

#include "spm.h"
#define  USAGE "usage: spm arrival_rate_(per sec) " \
               "pulse_length_(sec) source_capacity " \
               "number_of_pulses seed_(>0)\n"

unsigned long seed, init_seed;

main(int argc, char** argv) {
  if (argc != 6) {
    cerr << USAGE;
    return 0;
  }

  istrstream isar(argv[1], strlen(argv[1])+1);
  double arrival_rate;
  isar >> arrival_rate;
  istrstream ispl(argv[2], strlen(argv[2])+1);
  double pulse_length;
  ispl >> pulse_length;
  istrstream issc(argv[3], strlen(argv[3])+1);
  int capacity;
  issc >> capacity;
  istrstream ismc(argv[4], strlen(argv[4])+1);
  long max_requests;
  ismc >> max_requests;
  istrstream issd(argv[5], strlen(argv[5])+1);
  issd >> init_seed;
  seed = init_seed;

  if (!isar || !ispl || !issc || !ismc ||
```

```
       !issd || arrival_rate <= 0.0 ||
       pulse_length <= 0.0 || capacity < 1 ||
       max_requests <= 0L || seed <= 0) {
    cerr << USAGE;
    exit(0);
  }

  cout << "SPM Simulation Program   (M.C.Sinclair)\n\n";

  source spm(arrival_rate, pulse_length, capacity);

  spm.simul(max_requests);
  spm.print_out();

  exit(0);
}
```

The source.C source file
This file defines the seven member functions of the source class that were not
declared inline in spm.h. The constructor (source::source) sets the
arrival_rate, pulse_length and capacity to the values requested, as
well as zeroing the statistical data members. The source::req_arrival
function, which is invoked by the action member function of a request_arr
event, increments the number of requests (num_requests), invokes either
pul_start or req_blocked as appropriate, and then arranges to schedule a
new request_arr event after an appropriate interval. The
source::req_blocked function simply increments the count of the number of
blocked requested (num_req_blocked). The source::pul_start function
increments the count of the current number of pulses (num_pulses) and
schedules a pulse_end event after the appropriate interval (pulse_length).
The source::pul_end function, invoked by the action member of a
pulse_end event, simply decrements the current number of pulses. It should
perhaps be noted at this point that there are no true 'pulse start' or 'request blocked'
events—only member functions of class source—because they always occur at
the same time as a request_arr event, and so no separate event needs to be
scheduled. The source::simul function schedules the initial request_arr
event, and then repeatedly actions events until the requested number of pulse
requests has been generated. Finally, the source::print_out function outputs
the results of a run.

```
/*
  source.C
*/

#include "spm.h"

source::source(double ar, double pl, int c) {
  assert(ar > 0.0);
  assert(pl > 0.0);
  assert(c > 0);

  arrival_rate   = ar;
  pulse_length   = pl;
  capacity       = c;

  num_pulses     = 0;
  num_requests   = 0L;
  num_req_blocked = 0L;
}

void source::req_arrival(void) {
  assert(num_pulses >= 0);
  assert(num_requests >= 0L);
  assert(num_req_blocked >= 0L);

  num_requests++;

  if (!busy())
    pul_start();
  else
    req_blocked();

  double req_interval = expgen(arrival_rate);
  event* e =
    new request_arr(evq.time()+req_interval, this);
  evq.join(e);
}

void source::req_blocked(void) {
  assert(busy());
  assert(num_requests >= long(capacity));
  assert(num_req_blocked >= 0L);
```

```
  num_req_blocked++;
}
void source::pul_start(void) {
  assert(!busy());
  assert(num_pulses >= 0);
  assert(num_requests > 0L);
  assert(num_req_blocked >= 0L);

  num_pulses++;
  event* e =
    new pulse_end(evq.time()+pulse_length, this);
  evq.join(e);
}

void source::pul_end(void) {
  assert(num_pulses > 0);
  assert(num_requests > 0L);
  assert(num_req_blocked >= 0L);

  num_pulses--;
}

void source::simul(long max_requests) {
  assert(max_requests > 0L);

  double req_interval = expgen(arrival_rate);
  event* e =
    new request_arr(evq.time()+req_interval, this);
  evq.join(e);

  while (num_requests < max_requests)
    evq.action(); //simulate
}

void source::print_out(void) {
  cout << "Initial Seed             : "
       << init_seed << '\n';
  cout << "Arrival Rate             : "
       << arrival_rate << " per sec\n";
  cout << "Pulse Length             : "
       << pulse_length << " secs\n";
```

```
cout << "SPM Source Capacity      : "
     << capacity << '\n';
cout << "No. of Pulses Requested : "
     << num_requests << '\n';
cout << "No. of Pulses Lost       : "
     << num_req_blocked << '\n';
cout << "GOS                      : "
     << double(num_req_blocked)/
        double(num_requests) << '\n';
cout << "Total Simulation Time    : "
     << long(evq.time()) << " secs\n";
cout << "Final Seed               : "
     << seed << '\n';
}
```

The event_q.C source file

This file defines the four member functions of the event_q class not declared inline in spm.h. The constructor (event_q::event_q) initialises the object data members, whereas the destructor (event_q::~event_q) steps along the circularly-linked event queue using two event pointers (p and q) removing event objects and deleting them, thereby freeing all the dynamic memory used by this event_q object. The event_q::join function inserts an event object into the correct place in the event queue: at the beginning if the queue is empty; between two queued event objects (p and q) if appropriate; or right at the end. It also increments the size of the queue. Finally, the event_q::action function removes the first event from the queue, adjusting the latter as necessary, updates the simulation clock (q_time) to the time of the event, and then invokes the event (e->action();) before finally deleting it.

```
/*
   event_q.C
*/

#include "spm.h"

event_q evq;

event_q::event_q(void) {
   q_time = 0.0;
   last   = 0;
   size   = 0L;
}
```

```
event_q::~event_q(void) {
  if (!this || !last) return;

  assert(size > 0L); // event queue must be non empty

  event* p = last;
  do {
    event* q = p;
    p = p->next;
    delete q;
    size--;
  } while (p != last);

  last = (event *) 0;

  assert(size == 0L); // event queue must be empty
}

void event_q::join(event* e) {
  assert(size >= 0L);              // event queue size
                                   // must be positive
  assert(e);                       // event e must be
                                   // non void
  assert(e->e_time >= q_time);     // event time must not
                                   // be before current
                                   // simulation time
  if (!last)
    last = e->next = e;
  else {
    event* q = last;
    event* p = last->next;

    do {
      if (e->e_time < p->e_time)
        break;
      q = p;
      p = p->next;
    } while (p != last);

    if (e->e_time < p->e_time) {
      // e goes between q and p
      q->next = e;
```

```
        e->next = p;
    }
    else {
      // e goes after last
      e->next = last->next;
      last = last->next = e;
    }
  }
  size++;
}

void event_q::action(void) {
  assert(size > 0L); // event queue must be non empty
  assert(last);        // event last must be non void
   event* e   = last->next;

  assert(e);               // event e must be non void

  last->next = e->next;

  if (--size == 0L)
    last = (event *) 0;

  assert(e->e_time >= q_time);   // event time must not
                                 // be before current
                                 // simulation time
  q_time = e->e_time;
  e->action();
  delete e;
}
```

Assertions
Throughout the four files that make up the programme, assertions are used to make explicit assumptions and perform run-time checking of the implementation. For example, event_q::action contains assertions to check that the event queue is not empty and that the last pointer is valid before use; that the event pointer to the front of the quene (last->next) is non-void; and that the time of the event e is not before the current simulation time. Not all possible assertions are made, but rather a reasonable selection to enforce the more critical, otherwise implicit assumptions.

Results

To provide some sample results, the SPM simulation programme was executed ten times each, for an arrival rate (λ) of 200 pulse requests per second, and runs of total length 100,000 pulse requests, 500,000 pulse requests, and 1,000,000 pulse requests. The results are given in the table below, and the means ± the 98% confidence interval plotted on a graph for the different run lengths. The horizontal line on the graph represents the analytical result from the Erlang B formula. Note firstly the reduction in confidence interval with increasing length of the individual runs, and secondly the puzzling discrepancy between the analytical result and the confidence bounded experimental mean for the longest runs. This sort of discrepancy can occur for a variety of reasons in making such a comparison, and would merit investigation if it occurred in the course of research.

Table 3.5 Results of SPM simulation

	100k	500k	1M
1	0.009230	0.008250	0.008368
2	0.008780	0.008398	0.008387
3	0.008730	0.008382	0.008137
4	0.007690	0.008524	0.008226
5	0.009060	0.008568	0.008184
6	0.009400	0.008472	0.008346
7	0.007860	0.008578	0.008192
8	0.008390	0.008704	0.008236
9	0.008920	0.008502	0.008369
10	0.007650	0.009012	0.008144
Mean	0.008571	0.008539	0.008259
98% C.I.	0.000573	0.000185	0.000088

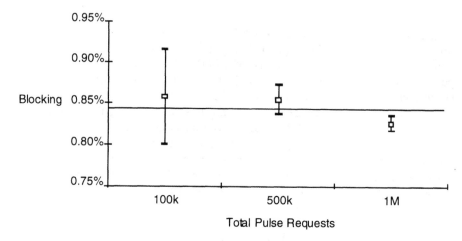

Figure 3.8   Results of subscriber's private meter simulation

3.6   References

1   FROST, V.S., LARUE, W.W. and SHANMUGAN, K.S.: 'Efficient
      techniques for the simulation of computer communications networks',
      IEEE J.Sel.Areas Commun., **SAC-6**, pp.146–157 (1988)

2   KUROSE, J.F. and MOUFTAH, T.: 'Computer aided modelling, analysis
      and design of communication networks' *IEEE J.Sel.Areas Commun.,*
      SAC-6, pp.130–145 (1988)

3   FISHMAN, G.S.: 'Principles of discrete-event digital simulation'
      Wiley (1978)

4   TAUSWORTHE, R.C.: 'Random numbers generated by linear
      recurrence modulo two', *Math Comput* ., **19**, pp.201–209 (1965)

5   KERNIGHAN, B.W. and RITCHIE, D.M.: 'The C programming language
      (Second Edition)', Prentice-Hall (1988)

6   KUMAR, D.: 'A novel approach to sequential simulation', *IEEE
      Software*, **3**, pp.25–33 (1986)

7   MOUFTAH, H.T. and STURGEON : 'Distributed discrete event simulation for communication networks', *IEEE J.Sel.Areas Commun.*, **SAC-8**, pp.1723–1734 (1990)

8   PHILLIPS, C.I. and CUTHBERT, L.G.: 'Concurrent discrete event-driven simulation tools', *IEEE J.Sel.Areas Commun*, **SAC-9**, pp.477–485 (1991)

9   LIPPMAN, S.B.: 'The C++ primer', Addison-Wesley (1989)

10   STROUSTRUP, B.: 'The C++ programming language (2nd Ed.)', Addison-Wesley (1991)

11   STROUSTRUP, B.: 'What is object-oriented programming?', *IEEE Software*, **5**, pp.10–20 (1988)

12   MEYER, B.: 'Object-oriented software construction', Prentice Hall (1988)

13   MORGAN, B.J.T.: 'Elements of simulation', Chapman and Hall (1984)

Chapter 4

Queuing systems

4.1 Service systems

The concept of queuing for service is familiar to most people. For example in a bank or post office, customers arrive and, if no service position is free, join a queuing system. In some cases, each service position has a separate queue and customers have to make a decision on which queue to join (usually the shortest one, which then moves forward at the slowest pace!). More enlightened organisations make arrangements for a single queue to form and the person at the head of the queue moves to the service position that first becomes free. This is a much fairer system than the multi-queue, although the mean waiting time is about the same, since the clerks do not work any faster. The reason is that the statistical smoothing arising from combining the queues greatly reduces the chance of very long waiting times.

The essential structure of a queuing system is shown in Figure 4.1. The system has a pool of potential customers. From time to time a member of this pool will enter a demand for service into the system. The system consists of a service facility containing one or more servers, each capable of serving one customer at a time, and usually also a queue where a customer may wait if he or she arrives when all of the servers are already busy. Each customer in the queue will eventually enter the service facility and as soon as the service is completed the customer leaves the system, to return to the pool.

There are many other examples of service systems. In the early manual telephone exchanges, when several subscribers attempted to make a call within a short interval the operators handled the calls as best they could and some subscribers had to wait in a queue. Most electro-mechanical exchanges made no provision for queues and a call that could not be handled immediately was assumed to be lost. However, it is convenient to regard such service systems as queuing systems with a maximum permitted queue length of zero. In fact in most practical systems a limit is placed on the maximum length of queue. In a packet network for example, packets entering a switching node are queued in a buffer store but only if there is storage space available. At times of very high traffic, all the storage space will be occupied and a new packet will have to be discarded. Other examples of queuing systems in information engineering occur in computer systems, where tasks for common resources for storage, printing etc. are queued and serviced in turn.

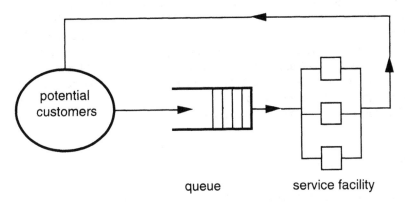

Figure 4.1 Structure of a queuing system

4.2 Performance assessment

A system designer should aim to ensure that the system will provide a service of an acceptable quality at an acceptable price. The system user will gauge the quality of the service according to various criteria, such as the system reliability, but in this chapter we shall be concerned exclusively with performance measures which quantify the rate of response of the system to its users' demands. Our intuition and experience tell us that the greater the load placed on a system's resources, the slower the response will be. The performance engineer seeks to quantify the relationship between system load and system performance.

Clearly, it is not possible to evaluate by direct measurement the performance of a system that has yet to be built. Even when the system is in operation, it may not provide mechanisms for the direct measurement of all performance measures of potential interest. Furthermore, if the measurement depends on the observation of a statistically rare event, a prohibitively long observation period might be required before any statistical significance could be attached to the result. For these reasons, the performance engineering of queuing systems relies heavily on mathematical modelling. First a model is constructed which embodies the essential characteristics of the queuing system. Subsequently, the likely performance of the queuing system may be deduced by a mathematical analysis of the behaviour of the model. In many cases, the analysis will only succeed if additional simplifying assumptions of a somewhat technical nature are made. Alternatively, a computer simulation of the system may be employed. The level of simplification may be less than that for the analysis model but it must be remembered that simulation will only provide sample statistics.

The performance measures that are used to evaluate a queuing system will depend on the application. The means and standard deviations of the waiting times may be useful but it is often more important to know the probability that a long waiting time will be exceeded. The economic aspects need to be considered. It is often possible to reduce waiting times by providing more service positions, and to reduce the probability of discard by increasing the buffer storage, but the increased complexity might then make the system uneconomic. Where human beings are connected to the system, some arbitrary decisions may be necessary to balance the annoyance factor produced by slow response times against the cost of the system.

4.3   Queuing models

The general description of a queuing system given above serves to identify the structure common to most systems of interest. As it stands, however, it is too vague to allow anything useful to be said about its performance. A queuing model is an idealised mathematical description of a queuing system, which specifies the following attributes in sufficient detail to determine the behaviour of the model to some desired level of detail:

(i)   the arrival pattern of incoming demands;
(ii)   the service requirements of individual customers;
(iii)   the capacity of the service facility for providing service;
(iv)   the queue discipline;
(v)   the capacity of the queue for holding jobs awaiting service.

The behaviour of a queuing system is illustrated in Figure 4.2. Customers arrive according to some measured or assumed process as shown in (a). Each customer has to be served and the service takes a time that depends on the nature of the system. Each instant indicated in (b) represents the completion of a service. If too many customers arrive in a short interval and/or if service times are too long, a queue is formed as shown in (c).

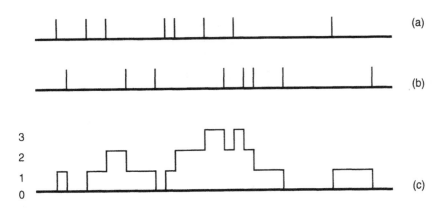

Figure 4.2 Behaviour of a queue

4.3.1 The arrival process

It is natural to describe the pattern of arrival of customers at a queuing system by reference to a time scale, $-\infty < t < \infty$. It is convenient to identify the present time with $t = 0$, and imagine that the queuing model is to be used to predict the future behaviour of the system during the period $0 \le t < \infty$. The instants at which successive customers enter the system will be denoted $t_1, t_2, t_3,...$, where $0 \le t_1 \le t_2 \le t_3 \le...$.

A service system for which the time of arrival of all future customers is already known at time $t = 0$ is said to have a deterministic arrival process. For example, a processor in a telephone exchange may be set up to distinguish between different types of tasks and service them according to a clocked scheduling system.

In many systems of practical interest, however, the arrival process is not deterministic, that is it is not possible to predict the future arrival instants. In this case it is convenient to model the unpredictable arrival pattern as a stochastic point process , that is to say, only the statistical properties of the sequence $\{t_1, t_2, t_3, ...\}$ are specified, and these are chosen to reflect the observed or predicted characteristics of the pattern of arrivals. The primary parameter of the arrival process is the average arrival rate, which is given the symbol λ. For instance, in Erlang's classical model of telephone traffic, it is assumed that the arrival pattern is completely random, that is, arrival instants are distributed uniformly in time, a notion which is formalised in the Poisson process defined in Chapter 2.

One factor which has an effect on the arrival process is the size of the customer population external to the service system. Clearly, the number of customers present in the system at time t determines the number of potential customers remaining in the pool, which in turn affects the arrival rate. For example, if a taxi service has only five taxis, when one taxi is already in communication with the base station over the mobile radio the arrival rate for new calls is reduced by a fifth. However, if the total number of potential customers is large this effect is usually slight, and in the limiting case of an infinite population, the arrival process can be assumed to be independent of the number in the queuing system.

Many of the queuing models that we shall study will have a general independent (GI) arrival process, that is, the interarrival times $T_j = (t_j - t_{j-1})$ will be assumed to be independent and identically distributed random variables. In this case the mean arrival rate is constant: $\lambda = 1/\overline{T}$, where $\overline{T}$ is the mean interarrival time. The Poisson process is a general independent process with negative exponential interarrival time distribution:

$$P[T_j \leq x] = 1 - e^{-\lambda x} \qquad | x \geq 0$$

$$(4.1)$$

Although it is usually justifiable to assume a constant arrival rate, there are some exceptions. For example, if a stored program controlled telephone exchange suffers a processor 'crash' such that all existing calls are lost, the call arrival rate will accelerate when the service is restored and subscribers attempt to reconnect their interrupted calls. Such cases are generally outside the scope of this book although the corresponding effect of an accelerating fault (arrival) rate is important in connection with reliability studies (see Chapter 8).

4.3.2   The service facility

In order for an arriving or waiting customer to begin service the customer has to enter the service facility. The service facility contains one or more servers, each capable of serving one customer at a time. In order to quantify the amount of service demanded by a customer, and the rate at which a server can provide service, a unit of work must be specified. The unit of work will depend on the service which the system provides. For instance, if the server is a digital transmission channel and the arriving customers are data messages or packets, the unit might be [bits] or [bytes], and the server will work at a rate measurable in bits per second. Alternatively, if the server is a CPU and the customers are executing programs, an appropriate work rate may be instructions per second.

The amount of service demanded by an arriving customer is often unpredictable. Moreover, it may depend in some way on the state of the system during the course of that customer's sojourn there, for example, on the time he spends waiting, or the level of congestion in the system, etc. However, in order to simplify the analysis, we shall restrict our attention to queuing models in which the amount of service demanded by each customer is fixed (but usually unknown) from the instant of his entry into the system. We shall also assume that each customer who has succeeded in entering the system, leaves it immediately after his service is completed, and (usually) not before.

In general the work rate of a server may also vary during the course of time, and may also depend on the state of the system. In addition the service facility may contain two or more servers working in parallel, but at differing rates. However, unless the contrary is stated, we shall assume that the service facility contains a fixed number (r) of identical servers, each of whom works at the same constant rate while busy and can remain idle only if there are no customers waiting for their service to begin. It follows from these assumptions that all service demands may be transformed unambiguously into service times by a simple scaling operation. Indeed from now on we shall usually measure all service demands in units of time. However, we shall need to consider service demands measured in work units later when we study networks of queues with state-dependent service rates.

In some queuing systems the service times are predictable, or very nearly so, for example, all equal to a constant. An example of such a system is a packet network which only admits constant length packets, such as an ATM network. Since the bit-rate over the multiplex is fixed, each packet will have the same transmission time. However, in many systems, individual service times are unpredictable, and it is possible to predict only their relative frequency, or probability distribution. In the queuing model we represent the service time of the jth arrival by a random variable, denoted S_j, (j = 1,2,3,...) In most cases, we shall assume general independent (GI) service demands. These random variables have common mean equal to the average service time S.

4.3.3   *Queue discipline*

Suppose that several customers are waiting in a queue, and a server becomes available. The queue discipline determines which of the waiting customers the free server will now select for service. Three possible simple rules of selection are:

(i)   First-come-first-served (FCFS), or first-in-first-out (FIFO)
(ii)   Last-come-first-served (LCFS), or last-in-first-out (LIFO)
(iii)   Random order of service (ROS), or service-in-random-order (SIRO)

FCFS gives priority to the customer who has been waiting longest. LCFS gives priority to the customer with the least waiting time, and is an appropriate discipline for modelling the behaviour of a stack in a processor. ROS gives no priorities at all: each customer waiting has an equal probability of being the next to enter service; it has been used to model the behaviour of the switchboard operators in a manual telephone exchange.

FCFS and LCFS assign priority to customers on the basis of arrival time. If the service time of every customer waiting were known, priority could be based on service time instead. For example, shortest-service-time-first (SSTF, or SIFO) is an ideal which CPU scheduling algorithms seek to attain, in order to minimise the average system response time. More generally, priority need not be based on the customer attributes so far discussed; the system designer may simply decide that some types of customer are more important than others and deserve better treatment. This situation is modelled by assigning a priority value from the range 1,2,3,...,P to each customer in the pool; when a server becomes free the waiting customer with the highest priority is selected for service.

In priority queues the queue discipline must dictate the action to be taken if a high priority customer arrives to find a customer of lower priority in service, when the server is faced with two alternatives. Either the service is completed for the low priority customer or the service is interrupted so that the high priority customer can be serviced immediately. The customer whose service is interrupted is said to be pre-empted by the high priority customer. When a pre-empted job returns to the service facility, the server may either resume working on the uncompleted portion of the original service time, or repeat the whole original service time, or even select a new service time by resampling from the service time distribution.

The round-robin scheduling algorithm is a pre-emptive resume queue discipline: each time a job reaches the head of the queue, it receives a fixed quantum of service, after which it returns to the back of the queue, or leaves the system if its service is now complete. In the limit as the quantum size shrinks to zero, each job in the system receives continuous service at a rate equal to the work rate of the server divided by the number of jobs present. Such algorithms are sometimes employed for access to the central processor unit in a computer.

A queue discipline is said to conserve work if it has no effect on the amount of service demanded by each customer and if no server is idle when there are customers waiting for service. In other words, a work-conserving queue discipline does not create or destroy work. All of the disciplines discussed so far are work-conserving, except those that enforce pre-emptive repeat, with or without resampling.

4.3.4   Finite capacity queues

The majority of queuing systems encountered in practice impose an upper limit on the total amount of work which is allowed to accumulate in the system at any time. In this case, we talk of finite capacity or finite buffer or limited waiting room queuing systems. For example, the total amount of message data (bits) which can be stored at a switching centre in a store-and-forward communications network is limited by the finite capacity of the buffer. The number of active jobs competing for the resources of a CPU is limited by the size of the main memory; thus an appropriate queuing model should limit the total number of jobs allowed to be in the system together.

When a queuing system becomes full to capacity, it will not accept any more jobs until it has spare capacity once more: in some systems the arrival process would be temporarily disabled by a control mechanism, e.g. flow control in a data network, or the job scheduling queue (JSQ) in a multiprogramming system; in others, jobs would continue to arrive but, on finding the system full, would leave immediately without receiving any service, i.e. they would be lost.

Just as a queue discipline may be exercised on the output of the queue as offered to the server(s), so a discipline may be applied to the input of the queue. Access to some of the positions at the back of the queue may depend on the nature of packets already in the queue. These are known as feedback controlled access queues. For example, in a packet system a packet may only be allowed access to one of the reserved positions at the rear end of the queue only if another packet from the same source is not already in the queue. This will ensure that if one source is generating packets at an exceptionally high rate, then packets from that source are more likely to be discarded than packets from one of the other sources (negative feedback control). Alternatively, admission to the queue may depend on at least one other packet from the same source being present in the queue (positive feedback control).

4.4 Kendall's notation

A shorthand notation, attributed to Kendall, is widely used to describe queuing models. It takes the general form:

$$A/B/C/m_1/m_2/Z$$

The six factors represent respectively arrival process/service process/number of servers/maximum capacity of queue(s)/population of users/service discipline. In the description of the arrival or the service process, the following symbols are used:

G: general (arbitrary) interarrival or service-time distribution
GI: general independent distribution
M: exponential distribution
D: deterministic (constant) distribution

The last factor (Z) represents the service discipline as described in Section 4.3.3. When the last three elements of Kendall's notation are omitted, it is usually understood that $m_1 = m_2 = \infty$ and that the queue discipline is FIFO.

4.5 Performance measures

The efficient management of a service system demands that its resources be fully utilised. On the other hand, if the utilisation is too high then the resulting congestion in the system may produce unacceptably long delays for the customers. The performance engineer must strike a balance between the conflicting requirements of increasing system efficiency and reducing customer inconvenience; hence the need to measure the performance of the system on both counts.

In the queuing model described above, which in the most general form that we shall study is the GI/GI/r queue, jobs of work of mean duration $\overline{S}$ are offered to the system at mean rate λ. It is important to note that from now on these and all other statistical properties of the input stream of traffic are assumed to be fixed at time $t = 0$ for all time, i.e. the workload is statistically stationary. This is justified in most practical cases since the period over which the system is observed is sufficiently short for changes in the level of traffic to be negligible and long enough for the statistical parameters to be estimated.

4.5.1   System performance measures

The throughput of a system is defined as the average number of jobs completed per unit time. For one server the maximum possible job completion rate $\mu = 1/\overline{S}$. Therefore, in a system of r identical servers working in parallel, the throughput cannot exceed $r\mu$ and we have

$$\text{throughput} = \min\{\lambda, r\mu\}.$$

$$(4.2)$$

Thus, the throughput is equal to the arrival rate λ as long as λ is less than the maximum service rate, μ, beyond which the throughput is saturated at the value $r\mu$.

The offered load, or traffic intensity, a is defined as

$$a = \lambda\overline{S}$$

$$(4.3)$$

This is a dimensionless quantity, but is often quoted in Erlangs; this should be interpreted as 'work-hours per hour' or 'work-seconds per second', etc. If the traffic intensity is greater than unity, work is coming in at a rate which is too great for one server to handle. If there are r parallel servers, then the system can handle a traffic intensity up to r; a system attempting to service a load exceeding r would experience ever increasing congestion, that is it would be unstable. The stability of an overloaded system ($a > r$) can be regained by introducing a mechanism for rejecting the excess work; of course, the load actually carried by the r servers would then be less than r.

The instantaneous level of congestion in a system can be gauged by the number of jobs waiting. Thus, the function of time defined as

$$Q(t) = \text{(the number of jobs waiting in the queue at time } t)$$

provides a continuous monitor of system congestion. Congestion can also be monitored through the functions defined as

$$N(t) \quad = \quad \text{(the number of jobs in the system at time t)}$$

and

$$N_S(t) \quad = \quad \text{(the number of servers busy at time t)}.$$

These three functions are related:

$$N(t) = Q(t) + N_S(t) \qquad (0 \le t < \infty)$$

$$(4.4)$$

In our stochastic model of the queuing system, $Q(t)$ is a random variable and $\{Q(t); 0 \le t < \infty\}$ is an integer-valued stochastic process in continuous time; so too are $\{N(t); 0 \le t < \infty\}$ and $\{N_S(t); 0 \le t < \infty\}$. Thus, the congestion in the model should be measured by means of the probability distributions of $Q(t)$, $N(t)$ and $N_S(t)$, and their summary statistics such as $E[Q(t)]$. Now, although the statistics of the incoming traffic stream are constant, it is to be expected that the statistics of the system will vary with time, and will also depend on the state of the system at time t = 0. However, for the class of queuing models that we shall study, it can be shown that, if the system is stable, after it has been in operation for a very long time indeed it will attain a condition known as statistical equilibrium; in this limiting condition the probability distributions of $N(t)$, $Q(t)$ and $N_S(t)$ assume stationary forms independent of t and of the initial state of the system. This independencemeans that analysis of the steady-state statistics of $N(t)$, $Q(t)$ and $N_S(t)$, that is those pertaining to equilibrium, is considerably less difficult than transient or time-dependent, analysis. As indicated earlier, steady-state analysis usually suffices for most practical purposes.

Thus, the carried load a' is defined as the steady-state average number of busy servers,

$$a' \quad = \quad \overline{N_S} \quad = \quad E[N_S(t)] \qquad \textit{in equilibrium.}$$

$$(4.5)$$

The difference $a - a'$ is a measure of the work rejected, or lost traffic. For systems which are stable without loss it can be shown that $a' = a$, as indeed we should expect.

The steady-state probability distributions and statistics of $N(t)$ and $Q(t)$, as well as providing a measure of the congestion in the system, are useful, for example, for determining the size of buffer required to hold waiting customers.

The occupancy or utilisation factor, ρ of a server is defined as the proportion of time that the server is busy, i.e. it is the load carried by that server. Clearly, the server occupancy cannot exceed unity. In a parallel system of r symmetrically loaded servers the server occupancy

$$\rho = a'/r .$$

(4.6)

For the GI/GI/r queue without loss, therefore,

$$\rho = \min\{ a/r , 1\}.$$

(4.7)

For a single server system without loss, the server occupancy, ρ and traffic intensity, a are the same if $a < 1$.

4.5.2 *Customer view of performance*

A customer who arrives to find that he cannot enter service immediately on account of all servers being busy is said to be blocked. Thus, an important measure of the congestion in the system is Π_B, the proportion of arrivals which are blocked. If blocked customers are cleared, that is we have a pure loss system, then Π_B is the probability of loss and is likely to be the only performance measure of interest to the customer. In telephone traffic parlance Π_B is known as the call congestion, as distinct from the time congestion P_B, which is defined as the proportion of time for which the system is in a blocking state. These two probabilities are not necessarily the same, unless the customers arrive at random, that is according to a Poisson process.

If blocked customers are delayed, the most important performance measure from the customers' viewpoint is the time spent in the queue or in the system. We define the waiting time W_j of the jth arrival to be the time that he spends waiting in the queue only, whereas the response time or sojourn time R_j is defined as the total time that he spends in the system. These random variables are related via the service time of customer j:

$$R_j = W_j + S_j \qquad (j = 1,2,3...)$$

(4.8)

Both $\{W_j, j = 1,2,...\}$ and $\{R_j, j = 1,2,...\}$ are continuous-valued random sequences, or stochastic processes in discrete time. Assuming that the system is stable, after the system has been in operation for a sufficiently long time to attain statistical equilibrium, the distributions of W_j and R_j assume stationary forms independent of j, t_j and the initial state of the system. The steady-state waiting time distribution $P[W \leq x]$ is probably the most important measure of the performance of a queuing system. Related single figure performance measures are the probability of delay, $\pi_B = P[W > 0]$, and the average delay $\overline{W} = E[W]$, that is the mean steady-state waiting time. The response time distribution $P[R \leq x]$ and its statistics may also be useful in some instances.

4.5.3   Calculation of performance measures

Thus far, we have described the rather general queuing model GI/GI/r, and a range of performance measures. It may appear as if all that is left to do is to provide general expressions for evaluating these. Unfortunately, this turns out to be extremely difficult. Indeed, very little is known about the behaviour of the GI/GI/r queue. Much more specific assumptions have to be made concerning the arrival process, service time distribution and number of servers, etc. before any performance measures can be calculated. We shall study some of the simpler soluble queues later. However, it is possible to give some general results here:

Stability
A queue whose workload is statistically stationary is stable if the occupancy of each server is less than unity ($\rho < 1$).

Arrivals see what departures leave behind them
Consider any queuing system in which customers arrive and are served separately, so that any realisation of the random processes $\{Q(t);\ 0 \leq t < \infty\}$ and $\{N(t);\ 0 \leq t < \infty\}$ is a step function with steps of unit magnitude only, either up or down. Clearly, the number of steps up from some level k can exceed the number of steps down to level k by at most one, and otherwise they must be equal. Consequently, in equilibrium the probability distribution of $Q(t)$ (or $N(t)$) just prior to an arrival instant is the same as that just after a departure instant.

Mean delay is an invariant
The mean delay is not affected by the queue discipline, provided that no work is lost or created inside the system and that the order of service is not based on a knowledge of the service times of individual customers. The average waiting time can be reduced if short jobs are given priority over long jobs, or increased by giving long jobs priority over short jobs, but arbitrarily altering the order of service merely reduces the delay for some jobs at the expense of others, leaving the mean unchanged. The waiting time distribution, however, is affected by the queue discipline.

4.6   Little's formula

Little's formula [1] is a simple yet extremely important identity, which amounts to an accurate and concise expression of the intuitive notion that the greater the number of jobs in a system, the longer time each job spends there. Little's formula is a general accounting identity of very wide application that we shall acknowledge by departing slightly from our established notation.

Let Σ stand for any queuing system or any part of a queuing system. Assuming Σ is in equilibrium, we define L as the mean number of jobs in Σ, λ as the average arrival rate into Σ, and W as the mean sojourn time spent in Σ. Little's formula relates the customer average W to the time averages λ and L as follows:

$$L = \lambda W.$$

(4.9)

Applying Little's formula to the whole queuing system, we find

$$\overline{N} = \lambda \overline{R}$$

(4.10)

where $\overline{N} = E[N(t)]$, $\overline{R} = E[R_j]$, both in equilibrium.

On the other hand, if Σ stands for the queue but does not include the service facility of the system, Little's formula becomes

$$\overline{Q} = \lambda \overline{W}$$

(4.11)

This equation can be established intuitively when the queue discipline is FCFS by observing that the average number of customers left behind in the queue when a customer enters service must equal the average number of arrivals during that customer's waiting time, which is λW. Similar reasoning may be used to establish the previous equation for single server queues only.

In fact, Little's formula is valid for any stable queuing system or subsystem in equilibrium, regardless of the arrival process, the service time distribution, the number of servers and the queue discipline, as long as it is work-conserving. A proof of this is given in Kleinrock [2] or in Koyayashi [3].

If Little's formula is applied to the service facility only of a queuing system with no loss, we find

$$\overline{N}_S = \lambda \overline{S}$$

$$(4.12)$$

In other words, the carried load is equal to the offered load ($a' = a$) for a pure delay system.

4.7 The Poisson arrival process

The Poisson arrival process has already been discussed in Chapter 2. We now develop the theory in relation to customers arriving at a queue. Consider a queuing system that commences operation at time $t = 0$. For all $t > 0$ we define $A(t)$ as the number of arrivals in the time interval $(0,t)$. $A(t)$ is an increasing step function of time. In a stochastic queuing model such as GI/GI/r, $A(t)$ is an integer-valued random variable for each $t > 0$, and $\{A(t); 0 \le t < \infty\}$ is a random counting process. Customers in a Poisson stream arrive singly at a rate λ which is uniform in time. In addition, the pattern of arrivals in a given time interval is independent of the pattern in any other disjoint (non-overlapping) interval. For example, arrivals in the interval (s,t) are independent of arrivals in $(0,s)$ and (t,∞). Hence, the Poisson process is sometimes called the completely random arrival process.

As shown in Chapter 2, the number of arrivals in any time interval t, has a Poisson distribution with mean λt and the interarrival time has an exponential distribution with mean $1/\lambda$. Furthermore, at an arbitrary time instant t_0 (the 'present'), the lapse of the (future) time until the next arrival has the same probability distribution as a complete interarrival time, regardless of the time which has already lapsed at t_0 since the most recent previous arrival. In other words, the Poisson process is memoryless.

4.7.1 Poisson arrivals see time averages

The delay experienced by a customer depends to some extent on the amount of work in the queuing system at the time of his arrival, and particularly when the queue discipline is FCFS. Thus, an important performance measure is the probability distribution of the number of customers in the system $N(t)$ when t coincides with an arrival instant.

We define the arriving customer's distribution at any time t as the conditional probability

$$\pi_k(t) = \lim_{h \to 0} P[N(t) = k \mid A(t,t+h)] \qquad k = 0,1,2,...,$$

$$(4.13)$$

where A(t,t+h) stands for the event: 'a customer arrives in the time interval (t,t+h)'. In general, the arriving customer's distribution and the unconditional probability distribution,

$$p_k(t) = P[N(t) = k] \qquad k = 0,1,2,...,$$

(4.14)

at the same time t are different. They are equal if, and only if, customers arrive according to a Poisson process, for then

$$
\pi_k(t) = \frac{P[N(t) = k \; \& \; A(t,t+h)]}{P[\, A(t,t+h)\,]}
$$

$$
= \frac{P[\, A(t,t+h) \mid N(t)=k]\; P[N(t) = k]}{P[\, A(t,t+h)\,]}
$$

$$
= p_k(t)
$$

(4.15)

since the probability of the event A(t,t+h) is independent of the state of the system at time t.

If the system is in equilibrium at time t, both $p_k(t)$ and $\pi_k(t)$ are independent of t. The steady-state distributions of N(t) will be denoted p_k and π_k respectively. The probability p_k is equal to the proportion of time for which there are k customers in the system. Thus, 'Poisson arrivals see time averages'. It follows from Section 4.5.3 that departing customers see time averages when they leave the system in equilibrium.

4.7.2 Exponential service time distribution

Consider a queuing system in which the service time distribution is negative exponential, i.e. for some $\mu > 0$,

$$P[S_j \le x] = 1 - e^{-\mu x} \qquad x \ge 0$$

From the reasoning given above, we know that this distribution is memoryless. Hence, we may deduce that, however long a customer has already been in service the residual service time has the same distribution as for a full service.

Suppose that a customer is in service at time t. The service will be completed before time t+h if the residual service time is less than h. Therefore, the probability that the customer completes before t+h is

$$1 - e^{-\mu h} = \mu h + o(h) \qquad \text{(for small h)}$$

Suppose now that altogether a total of k customers are in service at time t, and let D be the number of these who complete their service before time t+h. Then the conditional probability distribution of D is a binomial distribution Bin(k,p) with parameter

$$p = \mu h + o(h) \qquad \text{(for small h)}$$

(4.16)

Therefore, as $h \to 0$,

$$P[D = 1 \mid N_S(t) = k] = kp(1-p)^{k-1} = k\mu h + o(h),$$

$$P[D = 0 \mid N_S(t) = k] = (1-p)^k = 1 - k\mu h + o(h),$$

and $P[D \geq 2] = o(h).$

(4.17)

Thus, the service completion rate at time t is $N_S(t)m$.

Let us label the k customers in service at time t as 1,2,...,k in no particular order. Let X_i be the residual service time of customer i at time t. The customer with the smallest residual service time, which is $Y = \min\{X_1, X_2, ..., X_k\}$, will be the first of the k to complete his service. Since $P[X_i > x] = e^{-\mu x}$ for $x \geq 0$,

$$P[Y > x] = P[X_1 > x \ \& \ X_2 > x \ \& \ ... \ \& \ X_k > x]$$

$$= \prod_{i=1}^{k} P[X_i > x]$$

$$= e^{-k\mu x} \qquad\qquad x \geq 0$$

(4.18)

i.e. the time until the first completion occurs is exponentially distributed with mean $1/k\mu$.

4.8   The Markovian queue: M/M/r

Most queues are, in effect, birth and death processes. Customers arrive at random to join the queue (birth) and leave after service in a quasi-random process (death). Some fundamental aspects of birth and death processes have been introduced in Chapter 2. We now apply the theory to a specific queuing process.

Consider a queue with customers arriving in a Poisson process at rate λ, an exponential service time distribution of mean $1/\mu$ and an arbitrary number of servers. The system is an M/M/r queue with offered load $a = \lambda/\mu$. The queue discipline is assumed to be work-conserving, non-preemptive and independent of service time, but is otherwise arbitrary (e.g. FCFS, LCFS, or ROS). The M/M/r queue is analysed by recognising that the random counting process $\{N(t);\ t \geq 0\}$ is in this instance a birth and death process.

4.8.1   Steady-state distribution of $N(t)$

The number in system $N(t)$ increases by one at each arrival and decreases by one at each departure. It will be seen that the process satisfies the conditions for a birth and death process given in Chapter 2. Moreover, the memoryless property implies that these conditions are satisfied even if the queue discipline allows pre-emption (at arrival instants) with resume, repeat or resampling of subsequent service times.

With these birth and death coefficients, we find that

$$D_n = \frac{1}{n!} (\lambda/\mu)^n \qquad\qquad (n \leq r)$$

$$= \frac{1}{r!} \frac{1}{r^{n-r}} (\lambda/\mu)^n \qquad\qquad (n > r)$$

$$(4.19)$$

The tail of the sequence $\{D_n\}$ is in geometric progression with common ratio $\lambda/r\mu$. If $\lambda < r\mu$, the series ΣD_n has a finite sum

$$D = \sum_{j=0}^{r-1} \frac{a^j}{j!} + \frac{a^r}{r!} \frac{1}{1-\rho}$$

$$(4.20)$$

where $a = \lambda/\mu$ is the offered load and $\rho = \lambda/r\mu$ is the server occupancy.

Thus, if $a < r$ the equilibrium probability distribution of $N(t)$ is

$$P_n = \frac{1}{D} \frac{a^n}{n!} \qquad (n \le r)$$

$$= \frac{1}{D} \frac{r^r}{r!} \rho^n \qquad (n \ge r)$$

$$(4.21)$$

If $a \ge r$ then ΣD_n is infinite and $p_n = 0$ for all finite n, which indicates that equilibrium is impossible in this case.

A customer is blocked, that is delayed, if he arrives to find all r servers busy. In equilibrium the probability that all servers are busy is

$$P_B = \sum_{j=r}^{\infty} p_j = \frac{1}{D} \frac{a^r}{r!} \frac{1}{1-\rho}$$

$$(4.22)$$

Since Poisson arrivals see time-averages as explained above in Section 4.7.2, we may conclude that the probability that an arrival is blocked, $\pi_B = P_B$. This result is known as Erlang's delay formula:-

$$\pi_B = C(r,a) = \frac{\dfrac{a^r}{r!}}{(1-\rho) \displaystyle\sum_{i=0}^{r-1} \dfrac{a^i}{i!} + \dfrac{a^r}{r!}}$$

$$(4.23)$$

As the offered load a increases towards its upper limit r, the server occupancy r increases to 100% and the probability of delay $C(r,a)$ tends to unity.

Since $Q(t) = \max \{N(t) - r, 0\}$, the queue length distribution in equilibrium is given by

$$P[Q(t) = n] = C(r,a) (1 - \rho) \rho^n \qquad (n \ge 1)$$

$$= 1 - \rho C(r,a) \qquad (n = 0)$$

$$(4.24)$$

which is a modified form of the geometric distribution. This is also the distribution of the length of queue encountered by customers on arrival.

The conditional distribution of $Q(t)$, given that all servers are busy, is also geometric

$$P[Q(t) = k \mid N(t) \geq r] \quad = \quad (1 - \rho)\rho^k \qquad\qquad (k = 0,1,2,...)$$

(4.25)

It is interesting to observe that this distribution does not depend on the number of servers, that is an $M/M/r_1$ queue has the same conditional queue length distribution as an $M/M/r_2$ queue with the same server occupancy but with $r_1 \neq r_2$.

4.8.2   Average delay

The average queue length is given by

$$\bar{Q} \; = \; E[Q(t)] \; = \; C\,(r,a)\,\frac{\rho}{1-\rho}$$

(4.26)

Using Little's formula, we obtain the mean delay as

$$\bar{W} \; = \; \frac{C\,(r,a)}{(1-\rho)\;r\mu}$$

(4.27)

Examining these formulae we may draw the following extremely significant conclusion: as the server occupancy increases to 100%, the average number of jobs in the queue and the average delay both increase rapidly beyond all bounds. For this reason, it is usually unwise to operate a pure delay system at occupancies close to 100%, since a small increase in the average arrival rate may drive the system into an unstable condition. The effect is seen when the rate of arrival of vehicles on a section of road exceeds the capacity of the road; the queue rapidly builds up until action is taken to restrict the arrival of more vehicles.

> *Example: two computers on the same site currently transmit data messages to another site via separate data links, each of capacity L bits/sec., one dedicated to each computer. Each computer generates about the same amount of traffic. Data messages are transmitted complete. Messages which arrive*

> *when the appropriate link is busy are queued in a buffer. It*
> *has been proposed that each computer should be allowed to*
> *transmit on either link, using whichever happens to become*
> *available first: would this improve the service? Would it be*
> *better to replace the existing pair of links with a single link of*
> *capacity 2L?*

In the current system (a) we can model each dedicated link as an M/M/1 queue, assuming that the buffer is large enough to discount the possibility of overflow. Let λ be the message rate of each computer, and let $1/\mu$ be the mean transmission time. Using Erlang's delay formula, we find that the probability of delay $d_1 = \lambda/\mu$, and the mean delay in this case is

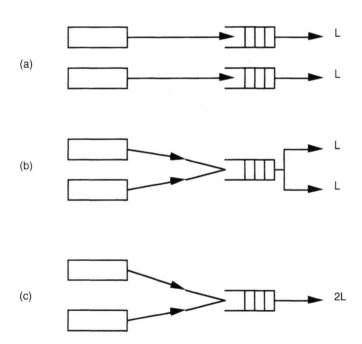

$$\overline{W}_1 = \frac{\lambda/\mu}{\mu - \lambda}$$

If the computers are able to share the links, as shown in (b) the system can be modelled as an M/M/2 queue with arrival rate 2λ. The probability of delay is now

$$d_2 = C(2, \frac{2\lambda}{\mu}) = \frac{2\lambda^2}{\mu(\lambda + \mu)}$$

$$= \frac{2\lambda}{\lambda + \mu} d_1$$

and the mean delay becomes

$$\overline{W}_2 = \frac{\lambda^2/\mu}{\mu^2 - \lambda^2}$$

$$= \frac{\lambda}{\lambda + \mu} \overline{W}_1$$

Since $\lambda < \mu$, we see that $d_2 < d_1$ and $\overline{W}_2 < 0.5\,\overline{W}_1$, i.e. the service should be substantially improved by pooling the work and sharing the resources.

We can model the single link of double capacity as an M/M/1 queue (c), with arrival rate 2λ and mean service time $1/2\mu$. This setup increases both the delay probability ($d_3 = d_1 > d_2$), and the mean waiting time, $\overline{W}_3 = 0.5\,\overline{W}_1 > \overline{W}_2$. On the other hand, if we consider the overall mean response times $\overline{R} = \overline{W} + S$,

$$\overline{R}_3 = \frac{1}{2(\mu - \lambda)}$$

whereas

$$\overline{R}_2 = \frac{\mu}{\mu^2 - \lambda^2}$$

$$= \frac{2\mu}{\lambda + \mu} \overline{R}_3$$

The mean response time with the double capacity link ($\overline{R}_3$) is evidently less than the response time for the separate links ($\overline{R}_2$) since $\lambda < \mu$.

4.8.3   *Waiting time distribution*

Although the calculation of average delay and queue length is of interest, in most systems the customers do not complain if the delay experienced is about average. It is the small proportion of customers who suffer exceptionally long delays that cause the problem. It is therefore necessary to calculate the probability distribution of customers' waiting times with the thought that we are mainly concerned with the tails of the distributions. It is to be expected that the probability distribution of customers' waiting times depends on the queue discipline. We shall calculate the equilibrium waiting time distribution for an M/M/r queue with FCFS scheduling only.

Suppose a particular job arrives to find $Q = j \geq 1$ jobs already waiting in the queue, an event whose probability is given above. Then, the new arrival enters the queue in position $j+1$ numbering back from the head of the queue, which is position 1. Each time one of the r jobs in service is completed, this particular job moves one position nearer to the head of the queue, from where it eventually enters service. Hence, his total waiting time, conditional on $Q = j$, is

$$W|_{Q=j} = Y_{j+1} + Y_j + \dots + Y_1$$

where Y_i is the time the job spends waiting in position i.

Now the residence time Y_i in position i is the time between two successive job completions with all r servers busy, except for Y_{j+1} which is the time between an arrival and the next completion. However, since service times are memoryless, all residence times have the same distribution as the time until the first completion out of r services in progress. It follows from Section 4.7.3 that Y_i has an exponential distribution with mean $1/r\mu$.

The sum of $j+1$ exponentially distributed random variables has an Erlang distribution of order $j+1$. (This should not be confused with Erlang's loss or delay probability). Hence, the conditional probability that the job waits longer than time t is

$$P[W > t \mid Q = j] = \int_{r\mu t}^{\infty} e^{-x} \frac{x^j}{j!} \, dx$$

$$(4.28)$$

Therefore, by the theorem of total probability, the probability that a delayed job is delayed longer than t is

$$P[W>t \mid W>0] = \sum_{j=0}^{\infty} P[W>t \mid Q=j \, \& \, W>0] \; P[Q=j \mid W>0]$$

$$= \sum_{j=0}^{\infty} (1-\rho)\rho^{j} \int_{r\mu t}^{\infty} e^{-x} \frac{x^{j}}{j!} \, dx$$

$$= (1-\rho) \int_{r\mu t}^{\infty} e^{-(1-\rho)x} \, dx$$

$$= \exp\{-(1-\rho) \, r\mu t\}$$

$$(4.29)$$

Finally, the unconditional probability that the waiting time exceeds t is

$$P[W>t] = P[W>t \mid W>0] \; P[W>0]$$

$$= C(r,a) \; \exp\{-(1-\rho)r\mu t\}$$

$$(4.30)$$

Thus, the waiting time distribution in the FCFS M/M/r queue is a mixture of a discrete probability at $t=0$ and an exponential distribution.

4.8.4 The M/M/1 queue

This is a particular case of the M/M/r queue and some simplification of the analysis is possible. A common example of a M/M/1 queue occurs when variable length packets are time multiplexed over a single digital link, so the theory is of some importance.

The state probabilities are given by

$$P_{n} = a^{n} P_{0} = a^{n}(1-a)$$

$$(4.31)$$

The mean number of customers in the system

$$E[N] = \sum_{n=0}^{\infty} n\, p_n = \frac{a}{1-a}$$

$$= \frac{\lambda}{\mu - \lambda}$$

$$(4.32)$$

Let R be the sojourn time (queuing plus service time). Suppose there are n other customers in the system when the observed customer arrives. Since the service times are memoryless, it will take an average time $1/\mu$ to complete service for the customer at the head of the line. For the remaining n processes, the mean service time will be n/μ.

Hence

$$E[R \mid n] = (n + 1)\frac{1}{\mu}$$

Therefore

$$E[R] = \sum_{n=0}^{\infty} (\mu + 1) \frac{1}{\mu}\, p_n$$

$$= \frac{E[N] + 1}{\mu}$$

$$= \frac{\dfrac{\lambda}{\mu - \lambda} + 1}{\mu} = \frac{1}{\mu - \lambda}$$

$$(4.33)$$

Hence $E[N] = \lambda\, E[R]$

which illustrates Little's result.

4.8.5   *Effect of finite buffer*

We now apply the restriction of a finite capacity queue as discussed in Section 4.8.1. Suppose the number of buffer positions is limited to K. If N(t) is the number of jobs in the system at time t then jobs arriving when N(t) are lost. N(t) is a birth and death process with state K the loss state (Figure 4.3).

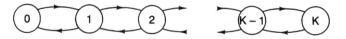

Figure 4.3 Markov representation of queue with finite buffer

Then $\qquad\qquad p_j \;=\; a^j p_0 \qquad\qquad\qquad 0 \le j \le K$

$\qquad\qquad\qquad\quad = 0 \qquad\qquad\qquad\qquad j > K$

The state probabilities form a truncated geometric progression so that

$$p_0 \;=\; \frac{1 - a}{1 - a^{K+1}}$$

and $\qquad\quad p_n \;=\; \frac{a^j(1-a)}{1 - a^{K+1}}$

$$(4.34)$$

The delay distribution may then be found as before. The probability of loss is found by putting $j = K$.

4.9   Queue with embedded Markov chain: M/GI/1

The analysis of the M/M/r queue was facilitated by the recognition that the number in system process $\{N(t); t \ge 0\}$ is a birth and death process in this instance. If either the interarrival time distribution or the service time distribution is not exponential, then $\{N(t); t \ge 0\}$ is not a birth and death process, or even a Markov process. However, for both GI/M/r and M/GI/1 queues it is possible to identify sequences of regeneration points in time which have the property that the values of $N(t)$ at these time instants define a Markov chain embedded in the process $\{N(t); t \ge 0\}$.

4.9.1 The Pollaczek-Khinchine formula

Consider a single server queue with customers arriving in a Poisson process of rate λ, and a general service time distribution $B(x) = P[S_j \le x]$. The queue discipline is work-conserving and non-pre-emptive, but is otherwise arbitrary, e.g. FCFS, LCFS or ROS. The regeneration points for this M/GI/1 queue are the time instants $\{\tau_1, \tau_2, \tau_3 \dots\}$ at which successive customers complete their service and depart from the system. Note that τ_j is the departure instant of the jth customer to be served, who is not necessarily the jth arrival. Thus, $N_j = N(\tau_j+)$ is the number of customers left behind in the system just after the jth departure.

Let X_j be the number of customers who arrive while the jth customer is in service. Clearly,

$$N_j = X_j \qquad \text{if } N_{j-1} = 0$$

$$= N_{j-1} + X_j - 1 \qquad \text{if } N_{j-1} > 0$$

It is convenient to write this as the single equation

$$N_j = N_{j-1} - U(N_{j-1}) + X_j$$

where we define $U(x) = 1$ if $x > 0$; $U(x) = 0$ if $x \le 0$.

Applying the z-transform to this equation, and noting that X_j is independent of N_{j-1} for the Poisson arrival process, we find that

$$E[z^{N_j}] = E[z^{N_{j-1} - U(N_{j-1})}] E[z^{X_j}]$$

$$(4.35)$$

Assuming the queue is stable, i.e. the server's occupancy $\rho = \lambda \overline{S} < 1$, in the limit as $j \to \infty$ the Markov chain will settle into statistical equilibrium and the probability distributions of N_{j-1} and N_j will be identical. Moreover, the steady-state distribution of N_j is the same as that of $N(t)$, because Poisson arrivals see time averages as explained in Section 4.7.2. Arrivals see what departing customers leave behind (see Section 4.5.3). Thus, for the M/GI/1 queue in equilibrium,

$$E[z^{N_{j-1} - U(N_{j-1})}] = p_0 + \sum_{n=1}^{\infty} p_n z^{n-1}$$

$$= p_0 + \frac{1}{z}(G_N(z) - p_0)$$

$$(4.36)$$

where $p_n = P[N(t) = n] = P[N_j = n]$, and $G_N(z)$ is the probability generating function (PGF) of $N(t)$ and N_j,

Now, the number of arrivals in a service time of length t has a Poisson distribution, that is

$$P[X_j = i \mid S_j = t] = e^{-\lambda t} \frac{(\lambda t)^i}{i!} \qquad i = 0, 1, 2, \ldots$$

Therefore the unconditional probability distribution is given by

$$P[X_j = i] = \int_0^{\infty} e^{-\lambda t} \frac{(\lambda t)^i}{i!} dB(t) \qquad i = 0, 1, 2, \ldots$$

$$(4.37)$$

Thus, the PGF of X_j is

$$G_x(z) = E[z^{X_j}] = \sum_{i=0}^{\infty} P[X_j = i] z^i$$

$$= \int_0^{\infty} e^{-\lambda t} e^{\lambda tz} dB(t)$$

$$= B^*(\lambda - \lambda z)$$

$$(4.38)$$

where

$$B^*(s) = \int_0^{\infty} e^{-st} dB(t)$$

is the Laplace transform of the service time distribution.

Substituting $G_N(z)$ and $G_X(z)$ into the recursive formula obtained earlier, and solving for $G_N(z)$ we find

$$G_N(z) = p_0 \frac{(z-1) B^*(\lambda - \lambda z)}{z - B^*(\lambda - \lambda z)}$$

(4.39)

where $p_0 = 1 - r$ is the probabilty that the system is empty. This result is known as the Pollaczek-Khinchine (PK) formula. It has to be inverted to obtain the steady-state distribution of N_j and $N(t)$. This can be accomplished only after the service time distribution $B(x)$ has been specified. However, the PK formula does allow the moments of $N(t)$ to be calculated for the general case. For example, by differentiating $G_N(z)$ and setting z equal to 1, after several applications of l'Hopital's theorem* we obtain the average number of customers in the system as

$$\overline{N} = G_N'(1) = \rho + \frac{\lambda^2 E[S^2]}{2(1-\rho)}$$

(4.40)

Since $\overline{N} = \rho + \overline{Q}$, an application of Little's formula reveals that the average waiting time in an M/GI/1 queue is given by

$$\overline{W} = \frac{\lambda E[S^2]}{2(1-\rho)}$$

(4.41)

This equation reveals that, like the M/M/r queue, the mean delay in an M/GI/1 queue tends to infinity as the server occupancy approaches unity. It also shows that the first moment of the waiting time W varies in proportion to the second moment of the arbitrary service time distribution $B(x)$. In other words, the variability of service times has a direct effect on waiting times: the more variable the service times, the greater the delay. For example, compare the cases where service times are (i) exponentially distributed, so that $E[S^2]=2(E[S])^2$, and (ii) constant, so that $E[S^2]=(E[S])^2$. It follows that the mean waiting time in the M/M/1 queue is twice

---

* L'Hopital's theorem states that if $\lim f = \lim g = 0$ then $\lim(f/g) = \lim(f'/g')$ provided the right hand side is well defined.

that in the M/D/1 queue offered the same average traffic. Furthermore, an M/GI/1 queue with finite mean service time, no matter how small, will have infinite mean delay if the variance of S is infinite.

4.9.2   Waiting time distribution in M/GI/1 FCFS queue

Suppose now that the queue discipline is FCFS. Then the jth customer to enter the system is also the jth customer to leave. Hence N_j must be equal to the number of Poisson arrivals $A(R_j)$ during the sojourn time R_j of the jth customer. Now the PGF of $A(t)$ for a time interval of given duration t is the PGF of a Poisson distribution of mean λt, that is

$$E[z^{A(t)}] = e^{-\lambda t(1-z)}$$

Hence, the PGF of $N_j = A(R_j)$ is

$$G_N(z) = E[z^{N_j}] = E[e^{-\lambda R_j(1-z)}]$$

$$(4.42)$$

Thus, the Laplace transform of the sojourn time distribution is given by

$$E[e^{-sR}] = G_N(1-s/\lambda) = \frac{(1-\rho)s B^*(s)}{s-\lambda + \lambda B^*(s)}$$

But $R = W+S$, where W and S are independent, and therefore

$$E[e^{-sR}] = E[e^{-sW}] B^*(s)$$

Hence, the Laplace transform of the waiting time distribution is given by

$$E[e^{-sW}] = \frac{(1-\rho)s}{s-\lambda + \lambda B^*(s)}$$

$$(4.43)$$

If the service time distribution has a rational Laplace transform, then this formula itself is a rational function of s, which can be inverted using the technique of partial fractions to evaluate the $P[W \leq x]$. For example, $B(x)$ is often assumed to be an Erlang or a hyperexponential distribution.

4.10 Slotted M/D/1 queues

The M/D/1 queue is usually dealt with as a special case of the M/GI/1 queue in which the service time is constant. It is assumed that service can be given to the customer at the head of the queue as soon as he or she arrives at that point. However, there is a particular case of a queue in which service can be started only at discrete intervals of time as determined by a clock. The intervals between clock pulses are assumed to be equal to the (fixed) service time. Then a customer, who may arrive at any time and finds the queue empty has to wait until the next clock pulse before service can be started. The system may be described as a 'slotted' M/D/1 queue. Examples of such behaviour occur in some computer systems and, as we shall see in Chapter 7, the asynchronous transfer mode (ATM) multiplexing system.

Up to now we have considered the behaviour of a queue as a continuous time process. However, in this particular case, it is sometimes advantageous to consider the behaviour of the queue as a discrete-time Markov process in which the condition of the queue is examined at the each clock pulse. This approach has the advantage that it facilitates the performance analysis of the system under conditions of continuous overload. Under such conditions, losses will occur when the capacity of the queue is exceeded as discussed in Section 4.8.5

For a Poisson arrival process, the number of customers (n) arriving at the queue during a service interval is given by the Poisson distribution:

$$P(n) = \lambda_n = \frac{L^n}{n!} e^{-L}$$

$$(4.44)$$

where L is the offered loading, that is $L = 1$ corresponds to 100% loading of the service system.

The notation λ_{n+} is used to indicate the probability of n or more customers arriving during the service interval, i.e.

$$\lambda_{n+} = \sum_{r=n}^{\infty} \frac{L^r}{r!} e^{-L}$$

$$(4.45)$$

The state diagram for the queue operation from state s is shown in Figure 4.5. During a service interval any number of customers may arrive at the queue and exactly one customer is served. If the number of arriving customers is two or more, the state is normally increased by the number of arrivals minus one, since one customer will have been served in the interval. If only one customer arrives the state remains stationary and if no customers arrive, the state is decreased by one. The exception arises if the capacity of the queue is exceeded and the model is restricted to state $k - 1$, where k is the number of positions in the queue. Note that the model can never attain state k since even if a customer is placed in the kth position in the queue, the performance of the service will ensure that he or she will have moved to the $(k-1)$th position at the instant the model is sampled.

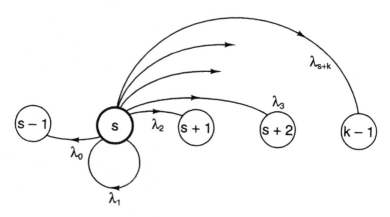

Figure 4.4 Possible transitions during a service interval

The concept of probability flow, introduced in Section 2.7.2, may be applied to discrete time systems The flow from one state to another is equal to the original state probability times the probability that a transition will take place to the second state at the end of the clock interval. Note that we are not concerned with what happens during the intervals between clock pulses. Then for statistical stability of a state, the sum of the probability flows into and out of that state must be equal to zero. The coefficients of the equations representing the stability of each state may then be expressed in matrix form

$$
M = \begin{bmatrix}
\lambda_0 + \lambda_1 - 1 & \lambda_0 & 0 & 0 & 0 & 0 & 0 & 0 & 0 \\
\lambda_2 & \lambda_1 - 1 & \lambda_{0} & 0 & 0 & 0 & 0 & 0 & 0 \\
\lambda_3 & \lambda_2 & \lambda_1 & \cdot & \cdot & \cdot & \cdot & 0 & 0 \\
\cdot & \cdot & \cdot & \cdot & \cdot & \cdot & \cdot & \cdot & \cdot \\
\cdot & \cdot & \cdot & \cdot & \cdot & \cdot & \cdot & \cdot & \cdot \\
\lambda_{(k-1)} & \lambda_{(k-2)} & \lambda_{(k-3)} & \cdot & \cdot & \cdot & \cdot & \lambda_1 - 1 & \lambda_0 \\
\lambda_{k+} & \lambda_{(k-1)+} & \lambda_{(k-2)+} & \cdot & \cdot & \cdot & \cdot & \lambda_{2+} & -\lambda_0
\end{bmatrix}
$$

$$(4.46)$$

For example, row 3 represents the coefficients of the equation for state 3, that is

$$\lambda_4 p_0 + \lambda_3 p_1 + \lambda_2 p_2 - (1 - \lambda_1)p_3 + \lambda_0 p_4 = 0$$

Note that instead of writing all the probability flows out of a state we have used

$$(1 - \lambda_1)p_s = (\lambda_0 + \lambda_2 + \lambda_3 + \ldots)p_s$$

since the sum of the transition probabilities out of a state is equal to unity.

The k equations subsumed in the matrix equation are not linearly independent and so the equations do not have unique solutions. It is therefore necessary to replace one of the equations by the normalising equation that the sum of the Markov states is equal to unity.

$$p_0 + p_1 + p_2 + \ldots + p_{k-1} = 1$$

If the equation corresponding to the last row of the matrix is chosen, this removes the necessity for calculating the sums of the Poisson probabilities. Computer techniques for the solution of equations expressed in matrix form are readily available [5], so that the state probabilities may easily be calculated. The delay distribution is then inferred from the state probabilities.

The condition for no service to take place during a service interval is for the model to be in state '0' and for no customers to arrive during the service interval. The probability of this occurring is $p_0 \exp(-L)$. The probability of a service being carried out during the service interval is then $1 - p_0 \exp(-L)$ and since the expected number of customers arriving during the interval is L, the proportion of customers turned away is

$$P_d = \frac{L - (1 - p_0 e^{-L})}{L}$$

(4.47)

4.11 Some useful approximations

For more general queues, with arbitrary distributions representing both interarrival times and service times, no readily computable exact expressions for even the mean waiting time are available. However, the need to have some idea of the behaviour of systems other than M/GI/1 (and GI/M/r) has lead to the generation of several bounds and approximations for W.

4.11.1   Kingman's heavy traffic approximation

As the occupancy of a GI/GI/1 server approaches one, the mean waiting time is given approximately by [4]

$$\overline{W} = \frac{\text{Var (T)} + \text{Var (S)}}{2\,\overline{T}\,(1-\rho)}$$

(4.48)

where Var(T) is the variance of the interarrival time, etc.

4.11.2   Whitt's approximation

The mean waiting time in a GI/GI/1 queue is approximately

$$\overline{W} = g \frac{\rho\,(C_T^2 + C_s^2)}{2\,(1-\rho)}\,\overline{S}$$

(4.49)

where $g = 1$ if $CT^2 \geq 1$, and

$$g = \exp\left[-\frac{2\,(1-\rho)}{3\rho}\,\frac{(1-C_T^2)^2}{C_T^2 + C_s^2}\right] \qquad \text{if } C_t^2 < 1$$

and where $C_T^2 = \text{Var(T)}/(E[T])^2,$ etc.

4.11.3   An approximation for GI/GI/r

The mean waiting time in a GI/GI/r queue is approximately

$$\overline{W} = \frac{C\,(r,a)}{2\,r\,(1-\rho)}\,(C_T^2 + C_s^2)\,\overline{S}$$

(4.50)

This formula agrees with the exact formulae for M/M/r and M/GI/1 queues.

4.12   References

1   LITTLE, J.D.C.: 'A proof of the queuing formula $L = \lambda W$', *Operations Research*, **9**, no.3, (1961) pp.383–387

2   KLEINROCK, L.: 'Queueing systems, volume I: theory', Wiley (1975)

3   KOBAYSHI, H.: 'Modeling and analysis: an introduction to system performance evaluation methodology', Addison-Wesley (1978)

4   KINGMAN, J.F.C.: 'On queues in which customers are served in random order', *Proceedings of the Cambridge Phil. Soc.,* **58**, (1962) pp.79–91

5   BAJPAI, A.C. *et al* : 'Engineering mathematics', Wiley (1974)

Chapter 5

Queuing networks

5.1 Networks of queues

In the basic queuing model discussed in Chapter 4, the customers contend with one another for the resources of a single service centre. In practice, computer and communications systems comprise several distinct service centres operating asynchronously and concurrently. For instance, in a multiprogrammed computer system, CPU's and I/O processors run in parallel. A packet switched network is another example, with the transmission link between each pair of switches acting as a separate service centre; separate queues of data packets waiting for transmission will form at each link. In order to study the problems of contention arising in systems like these, we represent each service centre as a single queue and model the whole system as a network of queues. As a particular job makes use of one service centre after another, we think of it as an object traversing a path through the network, visiting service centres one at a time.

Each node of a queuing network is a service system with its own service facility and queue. Jobs may enter the network at any node. If a job finds all servers busy when it enters a node, the job joins a queue in the usual way, to be selected for service at some later time. After completing its service at one node, a job may move to any other node, or reenter the same node for a subsequent bout of service, or leave the network.

A queuing network model specifies the exogenous arrival process entering fresh jobs at each node, the routing rule which determines the next node to be visited by a job which has just completed service in a node, the structure of the service facility and the queue discipline in each node, and the successive service demands made by a job at each node it visits. It may also be useful to divide jobs into different classes, each with its own routing rule and different service characteristics at each node, and even allow jobs to change class during their stay in the network.

We shall study a queuing network model of a somewhat restricted type, which is simple enough to enable basic performance measures to be evaluated exactly. Consider a network of N nodes, labelled 1,2,..., N (Figure 5.1). First, we assume that the routing rule is random: upon leaving a node, a job selects the next node it will visit, or elects to leave the network, by random selection. A job leaving node i goes next to node j with probability q_{ij}, j = 1,2,...,N, and leaves the network with probability q_{i0}. Of course, these routing probabilities must sum to unity:

$$\sum_{j=0}^{N} q_{ij} = 1 \qquad\qquad (i = 1,2,.....,N)$$

The internal transition probabilities are collected together into a routing matrix $\mathbf{Q}$ = $\{q_{ij}; 1 \le i, j \le N\}$.

Secondly, we assume that the ith node (i = 1,2,...,N) contains r_i identical GI servers working in parallel, that is the service times are independently distributed random variables with mean service time denoted $\overline{S}_i$; each time a job reenters a node its service demand is chosen again by random resampling. The exogenous arrival rate into node i is denoted λ_i. All the model parameters are assumed to be constant, that is time invariant.

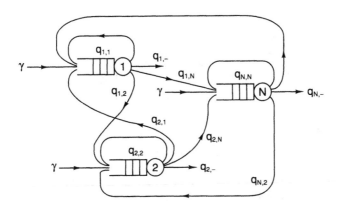

Figure   5.1 Simple network of queues

It is convenient to distinguish two kinds of queuing network, open and closed. In an open network, jobs enter the network from an infinite pool of customers outside, i.e. $\gamma_i > 0$ for at least one node i; and every job eventually leaves, i.e. from every node there is at least one route through the network to a node j where $q_{j0}>0$. Thus, the number of jobs in the network may vary. In contrast, a closed queuing network contains a fixed number $\mathbf{K}$ of jobs forever circulating from one node to another; no fresh jobs may enter the network, and none ever leave, that is $\gamma_i = q_{i0}$ = 0 for all i. Clearly, some queuing networks are neither open nor closed; they are said to be mixed, but we do not consider them further.

The formal mathematical analysis of the restricted type of queuing network considered here is similar, whether the network is open or closed. The distinction arises more from differences in behaviour and the methods used to evaluate their performance. Nevertheless, it is important to decide whether an open or a closed network model is appropriate for the system under investigation. For example, consider a very simple multiprogramming computer system, consisting of one CPU and one I/O device, which is subject to a maximum multiprogramming level of K.

Let us suppose that input jobs are placed in a job scheduling queue (JSQ) until a spare job slot is available. Once given a job slot, the job joins the queue for the CPU, and is eventually processed. The job then leaves the CPU, either to join the queue for the I/O device before returning to the CPU for further processing, or because it has terminated.

Under very light load, the degree of multiprogramming may hardly ever reach its upper limit K. In this case the job scheduling queue will always be empty and can be ignored; the system can then be modelled as an open queuing network of two nodes as shown in Figure 5.2.

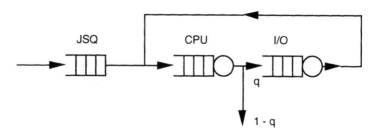

Figure 5.2 Open network model for light load

As the load increases the degree of multiprogramming may periodically reach its limit K. Under heavy load, the job scheduling queue will never empty, and whenever one of the K jobs in the system terminates another job will always be waiting to take its place. In this case it would be appropriate to model the system as a closed network (Figure 5.3).

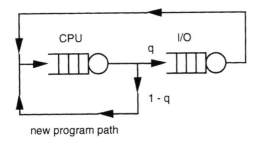

new program path

Figure 5.3 Closed network model for heavy load

Thus, the choice as to whether an open or closed model should be used may depend on the load. Store-and-forward communications networks are sometimes modelled as open networks; however, heavily loaded networks may limit the total number of messages in the network at the same time, e.g. by flow control, or token passing, in which case a closed model should be used. With computer systems, we are more likely to be interested in the performance under moderate to heavy load, where closed models are generally more appropriate.

5.2 Open networks of queues

Some basic conclusions may be drawn concerning the behaviour of open queuing networks without making any further assumptions.

5.2.1 Throughput

The input stream to node i is composed of the superposition of the exogenous arrival stream of rate γ_i and a proportion of each departure stream from nodes 1,2,...,N. If the throughput of node i is λ_i, and there is to be no net buildup of jobs in any node, we can equate λ_i with the overall arrival rate into node i:

$$\lambda_i = \gamma_i + \sum_{j=1}^{N} \lambda_j q_{ji} \qquad (i = 1, 2, ..., N)$$

$$(5.1)$$

Thus, if every node in the network is stable, this set of N linear equations, known as the traffic equations, must be satisfied by the set of N throughputs. Summing all N equations, we find that the total mean arrival rate into the network balances the mean network departure rate, that is the network throughput λ,

$$\sum_{i=1}^{N} \gamma_i = \sum_{j=1}^{N} \lambda_j q_{j0} = \lambda$$

(5.2)

In matrix notation, the row vectors $\underline{\lambda} = (\lambda_1, \lambda_2, ..., \lambda_N)$ and $\underline{\gamma} = (\gamma_1, \gamma_2, ..., \gamma_N)$ satisfy

$$\underline{\lambda}\ (I - Q) = \underline{\gamma}$$

(5.3)

If the network is open and the matrix $I - Q$ has an inverse, the traffic equations have a unique solution, which may be written as

$$\underline{\lambda}' = \underline{\gamma}\ (I - Q)^{-1}$$

(5.4)

However, the components of λ' can be interpreted as throughputs only if they do not exceed the maximum possible service completion rates at their respective nodes. In other words, the network model is stable only if the load offered is less than the number of servers at each service centre, that is $\lambda_i\ \overline{S}_i < r_i$ for $i = 1, 2, ..., N$. Clearly, the computation of the throughputs is meaningful only if these N conditions are all satisfied.

5.2.2   Visit count

The visit count V_i at node i is defined as the average number of occasions on which a job visits node i throughout its stay in the network. Since jobs enter the network at average rate λ, it follows that

$$\lambda_i = \lambda\ V_i$$

(5.5)

Thus, the visit counts for a given model may be calculated easily once the traffic equations have been solved. Visit count is a useful parameter in computer system performance evaluation, because it is a model parameter which can usually be estimated directly from measured data. For example, if 15,120 disk reads were recorded during a period when 720 interactions were processed, the visit count for the disk would be 21.

5.2.3 The average total service time

The average total service time σ_i that a job receives during all its visits to node i throughout its stay in the network is given by

$$\sigma_i = V_i \overline{S}_i$$

(5.6)

Like visit count, average total service time per job is a model parameter that can usually be estimated directly from measured data. For example, most FORTRAN compilers automatically return the total CPU processing time at the end of executing a program. The average service time per visit may then be estimated as σ_i/V_i. Since jobs enter the network at rate λ, the load offered to node i is equal to $\lambda\sigma_i$. Network stability demands that $\lambda\sigma_i < r_i$ for i = 1,..., N. Thus, the arrival rate λ_{sat} at which the network becomes saturated is determined by the node for which r_i/σ_i is least,

$$\lambda_{sat} = \min_{1 \le i \le N} \left(\frac{r_i}{\sigma_i} \right)$$

(5.7)

The mean total service time σ that a job receives from all nodes throughout its stay in the network may be calculated as

$$\sigma = \sum_{i=1}^{N} \sigma_i = \sum_{i=1}^{N} V_i \overline{S}_i = \frac{1}{\lambda} \sum_{i=1}^{N} \lambda_i \overline{S}_i$$

(5.8)

Suppose that a queuing network is in equilibrium at time t. At node i the number of jobs in service $N_{Si}(t)$ has mean value given by Little's formula, $\overline{N}_{Si} = E[N_{Si}(t)]$ $= \lambda_i \overline{S}_i$. Hence, the total number of jobs being served in the whole network $NS(t)$ is on average

$$\overline{N}_s = E[N_s(t)] = \sum_{i=1}^{N} \overline{N}_{S_i} = \sum_{i=1}^{N} \lambda_i \overline{S}_i = \lambda \sigma$$

(5.9)

Thus, it appears that Little's formula applies, in a somewhat subtle way, to the aggregate of all service facilities in an open network of queues.

5.3   Separable open queuing networks

Separable, or product-form, queuing network models are a subset of the general class of queuing network models, obtained by imposing further restrictions on the behaviour of the exogenous arrival processes and the service centres. In a separable network each service centre can be separated to some extent from the rest of the network for the purpose of evaluating its behaviour.

5.3.1   Jackson networks

Consider a queuing network of arbitrary topology on N nodes with routing matrix $\mathbf{Q} = \{q_{ij}\}$. Suppose that, for $i = 1,2,...,N$ the exogenous input to node i is a Poisson process of rate γ_i (possibly zero for each i, but not for all), which is constant in time and independent of the state of the network and the input processes to all other nodes. Suppose further that node i contains r_i identical servers and provision for an unlimited number of waiting jobs. The queue discipline is FCFS, and service times have an exponential distribution of mean $\overline{S}_i$ at node i. Let $N_i(t)$ be the number of jobs in node i at time t. Then, if the throughputs calculated by solving the traffic equations reveal that the network is stable ($\lambda_i \overline{S}_i < r_i$ for $i = 1,2,...,N$), it can be shown that as $t \to \infty$ the network will eventually attain an equilibrium in which $N_i(t)$ is independent of $N_j(t)$ if $i \neq j$. In other words the joint probability distribution of the random vector $\underline{N}(t) = (N_1(t),N_2(t), \ldots ,N_N(t))$ is separable, that is it factorises into the product form:

$$P(j_1,j_2, \ldots ,j_N) = P_1(j_1)P_2(j_2) \ldots P_N(j_N)$$

$$(j_i = 0,1,2, \ldots N ; 1 \leq i \leq N)$$

$$(5.10)$$

Moreover, the marginal distribution $P_i(j_i) = P[N_i(t) = j_i]$ is obtained by treating the ith node as an M/M/ri queue in isolation, with arrival rate λ_i.

This result was established by Jackson [1], who observed that the random vector $\underline{N}(t)$ defines an N-dimensional Markov process. The separable solution is proved by verifying that it satisfies the global balance equations which equate to zero the net rate of flow of probability into (or out of) any particular value of $\underline{N}(t)$. That it is the only solution follows from the general theory of Markov chains.

Jackson's theorem identifies a restricted class of open queuing networks, known as Jackson networks, in which individual service centres behave, at least as far as the number of jobs in each is concerned, as if all the jobs each centre receives arrive according to an independent Poisson process. This result is remarkable because in general the input streams are not Poisson; they are not even GI arrival processes.

5.3.2 *Performance measures*

Jackson's theorem enables the principle aspects of the time average behaviour of Jackson networks to be evaluated in a straightforward manner. For instance, the mean number of jobs at node i is given by

$$\overline{N}_i = E[N_i(t)] = \sum_{j=1}^{\infty} j\, P_i(j)$$

$$= a_i + C(r,a)\frac{\rho_i}{1-\rho_i}$$

(5.11)

where $a_i = \lambda_i \overline{S}_i$ is the traffic intensity, $\rho_i = a_i/r_i$ is the server occupancy, and $C(r,a)$ is Erlang's delay probability for the standard M/M/r queue.

We have seen that, for a single M/M/r queue in isolation, the waiting time distribution may be calculated from the distribution of $N(t)$, because Poisson arrivals see time averages. Unfortunately, similar reasoning cannot be applied to a single service centre in a general Jackson network, because the input stream is not Poisson. Consequently, it is not possible to calculate waiting time distributions for a general Jackson network. However, some customer averages can be calculated by the use of Little's formula. For example, the average sojourn time of a job at node i, that is the mean response time of node i, is given by

$$\overline{R}_i = \frac{\overline{N}_i}{\lambda_i} = \overline{S}_i + \frac{C(r_i,a_i)}{(1-\rho_i)r_i}\overline{S}_i$$

(5.12)

Clearly, the mean waiting time at node i is, therefore,

$$\overline{W}_i = \frac{C(r_i,a_i)}{(1-\rho_i)r_i}\overline{S}_i$$

(5.13)

The total number of jobs in the network is $N_T(t) = N_1(t)+N_2(t)+ \ldots +N_N(t)$. To find the mean response time of the network (that is, the time between the arrival and eventual departure of a job from the network) first evaluate the mean total number of jobs from

$$N_T = \sum_{i=1}^{N} N_i$$

$$(5.14)$$

The mean response time of the network may then be calculated using Little's formula,

$$R_T = N_T / \lambda$$

$$(5.15)$$

The validity of this result may be verified as follows:

$$\overline{R}_T = \sum_{i=1}^{N} V_i \overline{R}_i = \sum_{i=1}^{N} \frac{\lambda_i}{\lambda} \frac{N_i}{\lambda_i}$$

$$= \frac{1}{\lambda} \sum_{i=1}^{N} N_i$$

$$(5.16)$$

5.3.3 BCMP networks

In a joint paper, Baskett, Chandy, Muntz and Palacios [2] extended Jackson's theorem to cover somewhat more general queuing networks. They demonstrated that the distribution of $N(t)$ remains separable if the queue discipline at each service centre in the network is any one of the following:

FCFS	first-come-first-served
PS	processor sharing
LCFS-PR	last-come-first-served with pre-emptive resume
IS	infinite servers, or server-per-job

On arrival at an IS service centre, each job is assigned to a server and enters service immediately. There are always as many servers available as there are jobs in the system. Thus, in an IS node a job never has to wait. With this type of node we can model the collection of user terminals in a time-shared computer system, or the collection of CPU's in a multiprocessor system with shared memory.

In a FCFS node there is a single server whose work rate may depend on the number of jobs in the node, but the service demands must have an exponential distribution. In other words, the service completion rate at the node may be load dependent $\mu_i(N_i(t))$, but is otherwise independent of time. A node containing several $(r_i > 1)$ identical exponential servers, like a node in a Jackson network, can be modelled by setting

$$\mu_i(n) = \frac{min\,(n, r_i)}{\overline{S}_i}$$

In the other three cases the service demand distribution need not be exponential. The BCMP theorem [2] requires only that it have a rational Laplace transform, and it may be shown that the result can be extended to cover arbitrary distributions. Like FCFS, the work rate of the servers may depend on the number of jobs present. In its most general form, the BCMP theorem also allows jobs of different classes to have differing service demand distributions at the same (non-FCFS) node, but we shall only consider networks with a single job class.

For an open BCMP queuing network, like a Jackson network, the joint distribution of $\underline{N}(t)$ separates into the product of the marginals. The marginal distribution $P_i(j)$ at node i is obtained by treating $N_i(t)$ as a general birth and death process with birth coefficients $b_n = \lambda_i$ and death coefficients $d_n = \mu_i(n)$, independently of $Nj(t)$ where $i \ne j$. Of course, the usual M/M/r distribution still applies for a Jackson node. At a load independent node with a single server scheduled for FCFS, PS or LCFS-PR, the marginal distribution is geometric like that at an M/M/1 queue of occupancy $\rho = \lambda_i\,\overline{S}_i$, i.e.

$$P_i(j) = (1 - \rho)\,\rho^j \qquad\qquad (j = 0, 1, 2, \ldots.)$$

$$(5.17)$$

Clearly, the service completion rate at an IS node is load dependent. However, if every server works at the same constant rate while busy, the M/M/∞ model applies with traffic intensity $a = \lambda_i\,\overline{S}_i$, so that the marginal distribution is Poisson.

$$P_i(j) = e^{-a}\,\frac{a^j}{j!} \qquad\qquad (j = 0, 1, 2, \ldots.)$$

$$(5.18)$$

The response time at an IS node is equal to the service time because jobs never have to wait. In particular the mean response time

$$\overline{R}_i = \overline{S}_i$$

$$(5.19)$$

if node i is IS. At FCFS, PS and LCFS-PR nodes, $\overline{R}_i$ is calculated from $\overline{N}_i$ using Little's formula in the usual way.

5.4   Closed networks of queues

The behaviour of closed queuing networks differs fundamentally from that of open networks, as the following basic analysis reveals.

5.4.1   Relative throughput

In a closed queuing network, there are no exogenous arrivals, that is $\gamma_i = 0$ for all i. Therefore, the traffic equations are homogeneous. In matrix notation the throughput row vector λ satisfies

$$\lambda\,(\mathbf{I} - \mathbf{Q}) = \underline{0}\,.$$

(5.20)

Since $q_{i0} = 0$ for all i, the matrix $\mathbf{I} - \mathbf{Q}$ has zero row sums and, therefore, zero determinant. It follows that the traffic equations have an infinity of solutions, all equivalent up to a scale factor. For if $\underline{\lambda}^*$ is one solution, then $C\underline{\lambda}^*$ is another for any scalar C. Thus, the nodal throughputs of a closed network cannot be obtained from the traffic equations alone.

The actual throughput at each node depends not only on the maximum possible service completion rates there and at all other nodes, but also on the total number of jobs in the network K. However, given any solution $\underline{\lambda}^*$ to the traffic equations, we can obtain the relative throughput of any node, that is, relative to any other node j, for

$$\frac{\lambda_i[K]}{\lambda_j[K]} = \frac{\lambda_i^*}{\lambda_j^*}$$

(5.21)

5.4.2   Saturation point

If there is one job in a closed queuing network, there is no contention for resources and the response time of each service centre is equal to the service time. If the number of jobs K is greater than one, then both the throughput $\lambda_i[K]$ and the mean response time $\overline{R}_i[K]$ at node i depends on the contention for the resources of this and all other nodes. If K is increased, both $\lambda_i[K]$ and $\overline{R}_i[K]$ increase accordingly. As K increases from one upwards, the system becomes increasingly congested, but can never become unstable while K is finite (Figure 5.4).

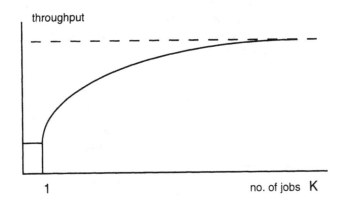

Figure 5.4 Saturation as number of jobs is increased

For given K the node most heavily loaded is that with the greatest server occupancy

$$\rho_i[K] = \lambda_i[K] \frac{\overline{S_i}}{r_i}$$

Although the actual occupancy is unknown, this node can be readily identified as the node with the greatest relative occupancy

$$\rho_i^* = \lambda_i^* \frac{\overline{S_i}}{r_i}$$

where each relative throughput λ_i^* is obtained from any one solution of the traffic equations. Since the relative occupancy is independent of K, the node so identified is the most heavily loaded at all values of K. As K increases this node is the first to become saturated, that is it is the 'bottleneck' of the network, because its occupancy approaches 100% faster than the occupancy of any other node.

Suppose node b is the bottleneck ($1 \leq b \leq N$). As K increases, the throughput of the bottleneck $\lambda_b[K]$ initially rises, until it saturates at the maximum service completion rate $r_b/\overline{S}_b$. Now the throughput at any other node i is proportional to $\lambda_b[K]$, for

$$\lambda_i[K] = \lambda_i^* \frac{\lambda_b[K]}{\lambda_b^*}$$

Thus, the bottleneck node effectively limits the throughput of all other nodes, whatever their service rates. Under these conditions, the system may be viewed as consisting of node b, with the rest of the network disregarded, as shown in Figure 5.5. It is futile to attempt to enhance a system's performance by increasing the capacity of any service centre other than the bottleneck.

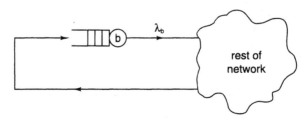

Figure 5.5 Effect of bottleneck on closed network

Applying Little's formula to the whole network, which contains exactly K jobs, we find that

$$K = \lambda_b[K] \left(\overline{R}_b[K] + \overline{R}_0[K] \right)$$

(5.22)

where $\overline{R}_0[K]$ is the mean response time of the rest of the network, that is, the average time a job spends away from node b between completing one service at node b and returning to node b for its next service thereafter. Rearranging the last equation we obtain a formula for the mean bottleneck response time,

$$\overline{R}_b[K[= \frac{\overline{S_b}}{r_b\, \rho_b[K]} K - \overline{R}_0[K]$$

(5.23)

Now, if K is sufficiently large, the bottleneck node is hardly ever empty, that is its occupancy $\rho_b[K] \cong 1$, and the average rate of entry of jobs into the rest of the network is constant at the maximum service completion rate $r_b/\overline{S}_b$. Consequently, the rest of the network behaves rather like an open network of queues, because most of the K jobs are queuing at the bottleneck, and, therefore, $\overline{R}_0[K]$ is independent of K. Thus, as $K \to \infty$ the mean response time at the bottleneck approaches the linear asymptote

$$\overline{R}_b[K] \approx \frac{\overline{S}_b}{r_b} K - \theta$$

where

$$\theta \;=\; \lim_{K \to \infty} \overline{R}_0 [K]$$

(5.24)

The effect is illustrated in Figure 5.6.

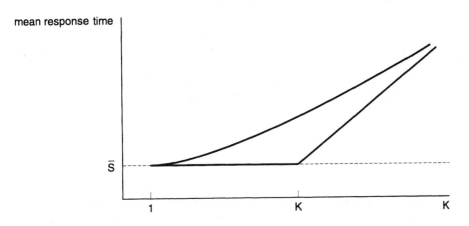

Figure 5.6 Mean response time at bottleneck

The value of K at which the linear asymptote intersects the horizontal asymptote $\overline{R}_b[1] = \overline{S}_b$, is given by

$$K^* \;=\; (1 + \theta / \overline{S}_b)\, r_b.$$

(5.25)

K* is called the saturation point of the network. The introduction of an additional job into a heavily loaded (K>>K*) closed queuing network effectively increases every job's response time by a proportion $1/r_b$ of the new job's service time.

Supposing it were possible to gradually increase the work rate of the servers in the bottleneck node b, doing so would initially raise the level at which throughput saturates in every node, i.e. it would raise the horizontal asymptote of each node's throughput curve. However, once the relative occupancy of node b fell below that at some other node β, say, continuing to decrease $\overline{S}_b$ would have little further effect. At this point node β would become the bottleneck of the system, and the asymptotic throughput of any node i would become $\lambda_i^* \, r_\beta /(\overline{S}_\beta \lambda_\beta^*)$. In other words, taking steps to increase the capacity of the primary bottleneck of a system may have disappointing results if there is a secondary bottleneck whose own throughput saturates at a level only slightly higher than that imposed by the primary bottleneck.The effect of improving the primary bottleneck on the mean response time curve is to displace the curve towards the linear asymptote of the secondary bottleneck, which intersects the horizontal axis at a higher saturation point K*.

5.5   Separable closed queuing networks

Consider a queuing network model on N nodes, which is like an open BCMP network in all respects, except that it is closed, i.e. $\gamma_i = q_{i0} = 0$ for $i = 1,2,...,N$. If there are K jobs in the network at all times, then the components of the random vector $\underline{N}(t) = (N_1(t), N_2(t), \dots N_N(t))$ must sum to K. Thus, the vector values which $\underline{N}(t)$ may assume form a finite set S(K,N) of size

$$| \ S(K,N) \ | \ = \ \left[\begin{array}{c} N+K-1 \\ K \end{array} \right] \ = \ \frac{(N+K-1)\,!}{(n-1)\,!\,K!}$$

(5.26)

It is clear that the components of $\underline{N}(t)$, i.e. the numbers of jobs in each node, cannot be independent in a closed queuing network. Nevertheless, the joint probability distribution of $\underline{N}(t)$ for a closed BCMP network in equilibrium is separable, that is it has the product form.

$$P\,(j_1, j_2, \dots, j_N) \ = \ \frac{1}{G\,[K]} \ \alpha_1(j_1)\, \alpha_2(j_2) \dots \alpha_N(j_N)$$

$$\text{for } j_i \geq 0; \qquad \sum_{i=1}^{N} j_i = K$$

(5.27)

However, the factor $a_i(j_i)$ is no longer the marginal distribution of $N_i(t)$. The term G[K] is a normalisation constant, that is it is chosen to ensure that the sum of the distribution over all possible values of $\underline{N}(t)$ is unity.

Let $\underline{\lambda}^* = (\lambda_1^*, \lambda_2^*, \dots, \lambda_N^*)$ be some non-zero solution of the traffic equations for the network. Then, for this $\underline{\lambda}^*$, the function $a_i(j)$ is defined as follows:

$$\alpha_i\,(0) \ = \ 1$$

and for $j \geq 1$,

$$\alpha_i(j) = \frac{(\lambda_i^*)^j}{\prod_{m=1}^{i} \mu_i(m)}$$

(5.28)

where $\mu_i(m)$ is the service completion rate of node i while $N_i(t) = m$.

Clearly, the normalisation constant G [K] is given by

$$G[K] = \sum_{j \in S(K,N)} \alpha_1(j_1) \, \alpha_2(j_2) \cdots \alpha_N(j_N)$$

(5.29)

However, it could prove extremely time consuming to compute G[K] directly from this summation. Fortunately, there is a more expedient method.

5.6 The convolution algorithm

5.6.1 Evaluating the normalisation constant

We define the set S(m,n) of vectors of integers for integers $m \geq 0$ and $n \geq 1$ as

$$S(m,n) = \{(j_1, j_2, \ldots, j_n) \,|\, \sum_{i=1}^{n} j_i = m \,;\, j_i \geq 0, \, i = 1, 2, \ldots, n\}$$

(5.30)

Clearly, S(K,N) is the set of all possible values for the state vector $\underline{N}(t)$.
    We also define the auxiliary function,

$$g_n(m) = \sum_{j \in S(m,n)} \alpha_1(j_1) \alpha_2(j_2) \cdots \alpha_n(j_n)$$

(5.31)

Thus, the normalisation constant $G[K] = g_N(K)$.

Now $g_1(m) = \alpha_1(m)$, and for $n > 1$, it is easily shown that

$$g_n(m) = \sum_{j=0}^{m} \alpha_n (m-j) \, g_{n-1}(j)$$

(5.32)

This equation reveals that the function $g_n(m)$ may be obtained recursively by forming the convolution of $g_{n-1}(m)$ with $a_n(j)$. It forms the basis of a numerical algorithm [3] devised for computing normalisation constants, known as the convolution algorithm.

In order to compute $g_n(m)$ for $n = 1, 2, \ldots\ldots, N$ and for $m = 1, 2, \ldots\ldots, K$, first the factors $a_i(j)$ must be computed for $j = 0,1, \ldots, K$ and $i = 1,2, \ldots, N$. The final (Nth) iteration of the algorithm yields not only $G[K]$, but also $G[1], G[2], \ldots\ldots, G[K-1]$. The implementation of the convolution algorithm requires a computer, but it is very simple to program.

The recursive formula (Equation 5.32) is valid for any BCMP node n, but if node n is load independent, that is it has a single server and $\mu_n(m)$ is a constant μ_n, then

$$\alpha_n(j) = (\rho^*)^j = \rho_n^* \cdot \alpha_n(j-1)$$

$$\text{where} \quad \rho_n^* = \lambda_n^* / \mu_n$$

(5.33)

and Equation 5.32 simplifies to become

$$g_n(m) = g_{n-1}(m) + \rho_n^* \, g_n(m-1)$$

(5.34)

Therefore, if node n contains a single server of fixed capacity with FCFS, PS or LCFS-PR scheduling, the Convolution Algorithm makes use of Equation 5.34 instead of Equation 5.32 to compute $g_n(m)$, $m = 1,2, \ldots, K$.

5.6.2   Computing performance measures

It can be shown that the throughput of node i is given by

$$\lambda_i [K[= \lambda_i^* \cdot \frac{G[K-1]}{G[K]}$$

(5.35)

Thus, the convolution algorithm must be executed completely in order to convert the relative throughputs obtained from the traffic equations into the true throughputs.

The convolution algorithm enables the probability $P(\underline{n})$ of any state $\underline{n} = (n_1, n_2, \ldots, n_N)$ of the network to be evaluated. The marginal distribution $P_i(j,K) = P[N_i(t)=j]$ at any node i can then be computed by summation of $P(\underline{n})$ over all states $\underline{n}$ where $n_i = j$,

$$P_i(j,K) = \sum_{\underline{n} \in S(K,N); \; n_i = j} P(n_1, n_2, \ldots, n_{i-1}, j, n_{i+1}, \ldots, n_N)$$

(5.36)

However, such a calculation would prove as laborious as computing $G[K]$ directly from its definition. Fortunately, this summation can be avoided by means of a second application of the convolution algorithm [3], this time to a network similar to the original network in all respects, except that node i is removed. The normalisation constants $G^i[k]$, $k = 1, 2, \ldots, K$, computed in this way enable the marginal distribution of the number of jobs at node i to be computed as

$$P_i(j, K) = \alpha_i(j) \frac{G^i[K-j]}{G[K]} \qquad j = 0, 1, 2, \ldots, K$$

(5.37)

Of course, this procedure must be repeated for each service centre whose marginal distribution is required.

The mean number of jobs in node i may be computed from the marginal distribution as

$$\overline{N}_i[K] = \sum_{j=1}^{K} j \, P_i(j, K)$$

(5.38)

Then the mean response time $\overline{R}_i[K]$ may be obtained using Little's formula. If the service centre has a load dependent service rate, this is the only way of calculating the mean response time. If the service rate is load independent, however, it is possible to compute these basic performance measures using only the initial application of the convolution algorithm to the original queuing network.

Suppose node i is load independent (LI), and recall that this means single server.

The server's occupancy $\rho_i[K]$ is then given by

$$\rho_i[K] = \rho_i^* \frac{G[K-1]}{G[K]}$$

(5.39)

The mean number of jobs in the service centre is given by

$$\overline{N}_i[K] = \frac{1}{G[K]} \sum_{j=1}^{K} (\rho_i^*)^j G[K-1]$$

(5.40)

Finally, the average sojourn time in node i, $\overline{R}_i[K]$, is obtained by an application of Little's formula

$$\overline{R}_i[K] = \frac{\overline{N}_i[K]}{\lambda_i[K]}$$

$$= \frac{1}{\lambda_i^* G[K-1]} \sum_{j=1}^{K} (\rho_i^*)^j G[K-j]$$

(5.41)

5.6.3   *Numerical problems with the convolution algorithm*

The convolution algorithm offers a theoretically exact method for evaluating the basic performance measures, including the marginal probability distributions of N(t), of a closed BCMP queuing network in which individual service centres may have load dependent service rates. With these two features, it is probably the most useful of all the available algorithms for solving closed networks. Of course, a computing machine is needed to actually implement the algorithm, and this in itself turns out to be the major source of problems; the normalisation constant G[K] may exceed the floating point arithmetic range of some machines for queuing networks with reasonable parameter values. In such cases the evaluation of G[K] may be impossible. For G[K] close to the smallest positive number in the floating point range, truncation errors can cause gross inaccuracy in the value determined by the convolution algorithm, giving rise to misleading results.

5.7   Mean value analysis

Mean value analysis (MVA) was developed by Reiser and Lavenberg [4] as an alternative to the convolution algorithm, to enable certain mean performance measures to be evaluated without the need to first compute the normalisation constant. MVA is a numerical algorithm which calculates recursively for $K = 1, 2, 3,$ ... the throughput $\lambda_i[K]$, the mean response time $\bar{R}_i[K]$ and the mean number of jobs $\bar{N}_i[K]$ at every node in a closed BCMP network of queues.

In principle, MVA can be modified to enable the marginal distributions $P_i(j,K)$ to be computed at the same time as the means. However, the modified version is unsuitable for machine computation, due to round-off errors. Unfortunately, this means that MVA is not generally suitable for application to BCMP networks with load dependent service centres, because the mean number of jobs at each such centre must be calculated from the distribution using Equation 5.38.

Mean value analysis is based on two principles:

(i)   Upon arriving at a service centre, a job finds that the distribution of the number of jobs already there is equal to the time average distribution for the centre with one job less in the whole network.

(ii)   Little's formula applies to the whole network, and to each service centre individually.

It can be shown that both principles are valid for closed BCMP queuing networks in equilibrium. Consider a closed BCMP network of N service centres, each containing either a single exponential server with load independent service rate, or more servers than there are jobs in the network (IS). Suppose there are K jobs in the network. Let $\underline{\Lambda} = (\Lambda_1, \Lambda_2, \ldots, \Lambda_N)$ be the unique relative throughput vector for which $\Lambda_1 = 1$. Then, if the throughput of node 1 were known, that of node i would be given by

$$\lambda_i[K] \;=\; \Lambda_i \, \lambda_1[K]$$

$$(5.42)$$

If node i is load independent, with FCFS scheduling, then an arriving job will remain in the service centre until the service of all jobs already there, plus its own service, is completed. The mean residual service time of the job in service on his arrival is equal to the mean full service time, because the exponential distribution is memoryless. It follows from principle (i) that an arriving job finds on average $\bar{N}_i[K-1]$ jobs in node i. Hence, the average sojourn time in node i is given by

$$\overline{R}_i[K[\; = \; \overline{S}_i \, (1 \, + \, \overline{N}_i[K-1])$$

(5.43)

Since the mean waiting time is invariant to queue discipline, this result is also valid for PS and LCFS-PR scheduling. The equation can be verified using Equations 5.40 and 5.41 relating $\overline{N}_i[K]$ and $\overline{R}_i[K]$ to the normalisation constant. If node i is an IS service centre then, whatever the value of K, the mean response time is simply

$$\overline{R}_i[K] \; = \; \overline{S}_i$$

(5.44)

Applying Little's formula to node i, we obtain an expression for the time average number of jobs there,

$$\overline{N}_i[K] \; = \; \lambda_i [K[\, \overline{R}_i[K]$$

(5.45)

Thus, the mean total number of jobs in the whole network is

$$K \; = \; \sum_{i=1}^{N} \overline{N}_i[K]$$

$$= \; \lambda_1[K] \sum_{i=1}^{N} \Lambda_i \, \overline{R}_i[K]$$

(5.46)

This equation can be rearranged to obtain an expression for the throughput

$$\lambda_1[K] \; = \; \frac{K}{\displaystyle\sum_{i=1}^{N} \Lambda_i \, \overline{R}_i[K]}$$

(5.47)

In the mean value analysis algorithm the four Equations 5.43 (or 5.44), 5.42 and 5.45 are used in roatation to enable $\overline{R}_i[K]$, $\lambda_i[K]$, and $\overline{N}_i[K]$ to be computed recursively for $K = 1,2,3, \ldots$, from the initial value $\overline{N}_i[0] = 0$.

5.8   Case studies

5.8.1 Central server model

Most contemporary multiaccess computer systems provide several time-sharing partitions in main memory, so that portions of several different users' programs can be stored simultaneously. Therefore, when an executing program requests an I/O operation, and its execution has to be temporarily suspended, one of the other programs in main memory can be (re-)started immediately. Thus, the CPU is utilised more effectively than it can be when there is only a single time-sharing partition in main memory.

Consider a multiprogrammed computer system with virtual memory subject to an upper limit of K_1 time-sharing partitions in main memory. The number of different programs in main memory, known as the level or degree of multiprogramming, may vary over time but can never exceed K_1. Usually, the number of terminals connected to the system, K_0 say, is substantially greater than K_1, and when there are more than K_1 users' requests to be processed, some of them have to wait in a job scheduling queue (JSQ) until allocated a region of main memory. Suppose the system has a single processor and several input/output (I/O) devices, where each I/O device is used either for demand paging or file transfer. For example, there may be a single paging drum with several logical sectors, and several discs for file transfer. A program continues execution until either its time-slice expires, or it page faults or issues a file transfer request, whichever occurs first. An I/O request causes the program execution to be suspended and one of the other programs in main memory (re-)starts execution. Time-slice expiry causes the job to be returned to the back of the JSQ, and its region of main memory is reallocated to a job waiting in the JSQ.

In this multiprogramming system, the scheduling and resource allocation are performed at two levels. First, in the outer level, up to K_0 jobs contend for a partition in main memory. Second, in the inner level, up to K_1 jobs contend for the subsystem comprising the CPU and I/O devices. A job being 'served' in the outer level may actually be waiting in a queue in the inner level. At first sight, it is difficult to see how both levels can be represented in one queuing network model. While a multiprogrammed system is operating under heavy demand, there is always a program waiting in the JSQ to be allocated a region of main memory. Therefore, whenever a running program completes execution, or exhausts its current time-slice, it is immediately replaced. Hence, the multiprogramming level is constant at K_1. Thus, under heavy demand the contention for the CPU and the I/O devices among the jobs allocated main memory can be represented by a closed network of queues known as the central server model (Figure 5.7).

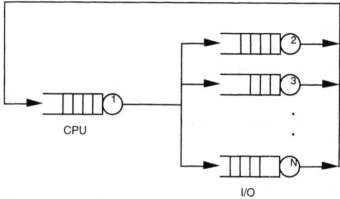

Figure 5.7 Central server model

Service centre 1 represents the CPU: it contains a single server scheduled for first-come first-served (FCFS) operation. Service centres 2 to N represent the I/O devices, i.e. the separate logical sectors of the paging drum, and the disk channels. These devices are scheduled for FCFS. The network contains K_1 jobs which represent whichever K_1 programs currently hold a region of main memory. Note that CPU utilisation and similar measures of performance at the inner level are not affected by the fact that each user program in main memory is periodically replaced by another. The jobs circulate endlessly from the CPU node to one or other I/O device, and then back to the CPU. The routing probabilities are chosen to represent the relative frequency with which the various I/O devices are accessed. The CPU and the I/O devices must be assumed to have exponential service time distributions. Then the central server model is a closed BCMP queuing network, and performance measures such as CPU or I/O utilisation, queue length distributions, etc., can be evaluated using the standard numerical algorithms discussed earlier.

The basic central server model does not distinguish the completion of a program's execution from other causes of I/O operations such as page faults, file transfer requests or the end of a job's time-slice. As it stands, therefore, it does not enable the throughput or mean response time of the computer system to be calculated. However, since the average time for which a program executes between making successive I/O requests, $\overline{S}_1$, is known, the proportion of all I/O requests which are caused by programs completing can be determined as

$$q_0 = \frac{\overline{S}}{t_p}$$

(5.48)

where t_p is the mean total CPU time required to execute a program to completion.

The termination of a program can then be represented, in a modified central server network (Figure 5.8) by a job re-entering service centre 1 immediately upon completing a service there. Thus, in the modified central server network a proportion q_0 of jobs leaving node 1 take the feed-back route labelled 'new program path'. Clearly, the throughput of the computer system, that is the average rate of completion of users' requests, is given by

$$q_0 \lambda_1(K_1)$$

(5.49)

where $\lambda_1(K_1)$ is the throughput of the CPU node 1.

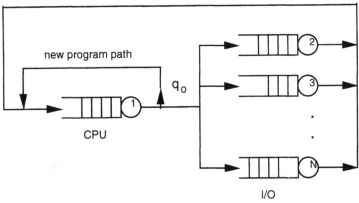

Figure 5.8 Modified central server model

5.8.2 Multiprocessor systems

In a multiprocessor system, program execution is shared among several processors operating concurrently, that is in parallel. The performance analysis of multiprocessor systems must take into account the possible contention among the processors for access to the main memory, which holds code and data for the executing programs. To reduce the memory contention, code and data are usually distributed over several store blocks. Different processors may access separate store blocks concurrently, but each block can deal with only one request for memory transfer at a time. Thus, a queue of waiting requests may arise at any store block. While a processor is waiting for memory transfer, it is doing no useful work.

Consider a system of K processors and $N-1$ store blocks as shown in Figure 5.9. Suppose it is operating under heavy load. Thus, the processors are either busy executing programs, or idle waiting for access to, or accessing, a store block. They can be represented by K jobs circulating endlessly around a closed queuing network on N nodes, whose topology is similar to the central server model. However, the central server here is an IS node, representing the processors which are executing code.

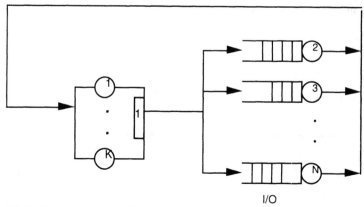

I/O

Figure 5.9 Multiprocessor model

The other nodes are FCFS service centres representing the N−1 store blocks. The random routing probabilities are determined by the relative frequency with which each block is accessed, and are assumed to be the same for all processors. Access times are fairly constant for random access memory, but it is necessary to assume exponential service times at the FCFS nodes. Then the network is a BCMP network and the usual performance measures, including the mean response time $\overline{R}_i$ at each node i, can be calculated using the standard numerical methods.

If

$$\overline{W}_i \;=\; \overline{R}_i - \overline{S}_i$$

is the mean time a processor waits for access to store block i (= 2,3, . . . ,N), then in one cycle the average time a processor wastes due to contention for memory is

$$W \;=\; \sum_{i=2}^{N} q_{1i}\, \overline{W}_i$$

(5.50)

Since the mean cycle time is

$$C = \bar{R}_1 + \sum_{i=2}^{N} q_{1i} \bar{R}_i$$

$$(5.51)$$

memory contention causes each processor to waste a proportion W/C of its potential processing time. Hence, the effective processing power of the system is equivalent to

$$\frac{K(C-W)}{C} \quad \text{processors}$$

5.9 References

1   JACKSON, J.R.: 'Networks of waiting lines', *Operations Research*, **5**, (1957) pp.518–521

2   BASKETT, F., CHANDY, K.M., MUNTZ, R.R. and PALACIOS-GOMEZ, F.: 'Open, closed and mixed networks of queues with different classes of customers', *Journal of the ACM*, **16**, (1975) pp.527–531

3   BUZEN, J.P.: 'Computational algorithms for closed queuing networks with exponential servers', *Communications of the ACM*, **16**, (1973)

4   REISER, M. and LAVENBERG: 'Mean value analysis of closed multichain queuing networks', *Journal of the ACM*, **27**, (1980) pp.313–322

Chapter 6

Switched systems

6.1 Objectives of traffic engineering

The objective of traffic engineering for, say, a telephone exchange is to produce a statistical design so that the complexity is handled in a nondeterministic way. The dimensioning of the component parts of the exchange is related to the required traffic capacity so as to meet given performance criteria such as loss probability, response time, etc.

In order to carry out such a design, we need to be able to calculate the traffic capacity of simple circuit groups, taking into account the stochastic properties of traffic, appropriate control procedures and the desired performance targets. Thus, by considering connection topology and the interaction of simple groups, we will be able to calculate the traffic capacity of more complex switches and hence derive design principles for both switches and networks.

It is of course possible to conceive of a system in which all demands receive immediate service. Such a system is said to be non-blocking. However, it would not be practicable to design a complete non-blocking network although the concept can be used for parts of the network. For a practical system, the inevitable blocking will result in loss, delay or diversion to an alternative route, or any combination of these. The design targets will usually be set as loss probability and/or delay probability. In some cases, it will be sufficient to define the mean delay, but in others it will be necessary to employ other statistics such as the tail of the delay probability distribution. Complaints are not made by the many customers who suffer delays close to the mean, but by the few who are kept waiting for much longer. Furthermore, even if the design of an exchange is such as to give satisfactory service under normal traffic conditions, the performance when the traffic is increased slightly may be such as to lead to a high level of customer complaints. It is therefore often necessary to impose some form of control on the overload behaviour.

Allowance should also be made for growth in the system. This entails forecasting the requirements for some time in advance. Although many techniques such as extrapolation, comparison with similar situations, knowledge of economic factors, etc., are available, there is inevitably a degree of uncertainty in the estimates. Hence it is doubly important to consider the effect of overload. The planning period for which equipment is installed to meet the next stage in growth depends on a combination of technical and economic factors discussed in works on telecommunications economics [2, 3].

6.2 Traffic sources and properties

Consider a simple concentrating switch such as shown in Figure 6.1. The number of sources is n and the number of circuits in the outgoing trunk group N. Since it is a concentrating switch, $n \gg N$. The sources are assumed to be independent and the mean traffic per source is a. It follows that $a \ll A$, where the latter is the traffic offered to the trunk group.

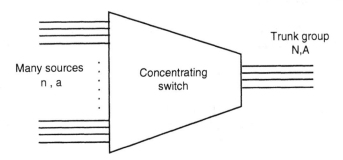

Figure 6.1 A simple concentrating switch

The traffic can then be modelled as a random process, that is a random variable going on in time. The randomness of the originating traffic has two aspects, the arrival process and the holding times. The circuit occupancy depends on these factors as shown in Figure 6.2.

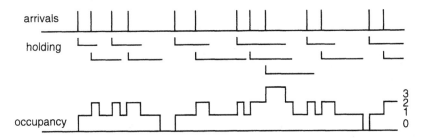

Figure 6.2 Distribution of circuit occupancy

This shows the arrival times as vertical lines, the holding times as horizontal lines, and the circuit occupancy stepping up and down as calls arrive and complete, respectively.

6.3   Modelling traffic sources

Essentially, traffic arrivals are modelled by independent random events with uniform probability density as explained in Chapter 2. Important parameters of the source characteristics which affect the network performance are the distribution of the inter-arrival times, the number of arrivals in a given time interval and the holding time for each arrival [5, 6]. In the following sections each of these parameters are modelled, assuming that the arrival distribution is uniform, and events are independent of each other.

6.3.1   Inter-arrival times

If independent random events with a uniform probability density function have a constant rate of λ, then the probability of only one arrival in a short time dt is

$$\text{prob[event in } (t, \, t+dt)] \; = \; \lambda \, dt$$

Suppose with probability $p_0(t)$ that no arrival has occurred in the time interval of $(0, t)$. Given that the probability of one arrival in $(t, \, t+dt)$ is λdt, and the probability of more than one is vanishingly small, then the probability of having no arrivals in this period is $1-\lambda dt$. Thus

$$p_0(t+dt) = p_0(t)(1-\lambda dt)$$

or

$$\frac{p_0(t+dt)-p_0(t)}{dt} = -\lambda \, p_0(t)$$

and hence

$$\frac{dp_0(t)}{dt} = -\lambda p_0(t)$$

$$(6.1)$$

Given the (obvious) boundary condition that $p_0(0) = 1$, the solution to this differential equation is an exponential function

$$p_0(t) = e^{-\lambda t}$$

$$(6.2)$$

To find the inter-arrival density distribution, $f(t)$, note from Chapter 2 that the probability of an event in an interval dt is equal to the integral of its density function in this interval. For a very short time dt, then

$$p(t < \tau \le t + dt) \equiv f(t)dt$$

$$(6.3)$$

$$f(t)dt = prob[\text{no event in } (0,t)] \cdot prob[\text{event in } (t, t+dt)]$$

which is

$$f(t)dt = p_0(t)\lambda dt$$

$$(6.4)$$

or

$$f(t) = p_0(t)\lambda$$

and substituting for $p_0(t)$ from Equation 6.2,

$$f(t) = \lambda e^{-\lambda t}$$

$$(6.5)$$

Thus for arrivals with uniform density and constant arrival rate, the probability density function of the inter-arrival times is a negative exponential function. Note that the mean value of this distribution is the inverse of the arrival rate, $1/\lambda$.

6.3.2 *Number of arrivals*

Consider again random arrivals with a constant rate of λ, having a uniform distribution. However, this time we assume there are r arrivals in the time interval $(0, t)$ with a probability of $p_r(t)$. If we increment the time interval by dt, and still expect to have r arrivals in this new time interval of $(0, t+dt)$, then either there should be $r-1$ arrivals in $(0, t)$ and one arrival in $(t, t+dt)$, or there should be r arrivals in $(0,t)$ and no arrival in $(t, t+dt)$, as shown in Figure 6.3

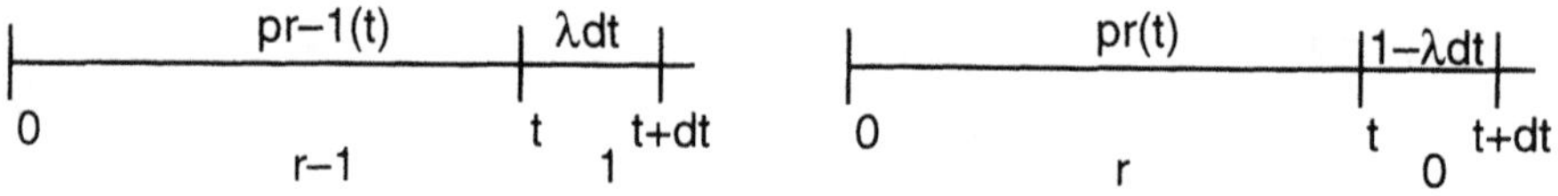

Figure 6.3   Probability of arrival

Thus

$$p_r(t+dt) = p_{r-1}(t)\lambda dt + p_r(t)(1-\lambda dt)$$

or

$$\frac{p_r(t+dt)-p_r(t)}{dt} = \lambda p_{r-1}(t) - \lambda p_r(t)$$

$$(6.6)$$

Using a generating function, as defined in Chapter 2,

$$F(z,t) = \sum_r p_r(t)z^r$$

$$\frac{d}{dt}F(z,t) = \sum_r \frac{d}{dt}p_r(t)z^r$$

$$= \lambda\sum_r p_{r-1}(t)z^r - \lambda\sum_r p_r(t)z^r$$

$$(6.7)$$

Substituting $u = r-1$ in the first summation, then

$$\frac{d}{dt}F(z,t) = \lambda z\sum_u p_u(t)z^u - \lambda\sum_r p_r(t)z^r$$

$$(6.8)$$

and using the generating function notation, $F(z, t)$, then

$$\frac{d}{dt}F(z,t) = \lambda zF(z,t) - \lambda F(z,t)$$

or

$$\frac{d}{dt}F(z,t) = (\lambda z - \lambda)F(z,t)$$

$$(6.9)$$

Given the boundary condition $F(z,0)=1$ (as if $t=0$, then $p_0(0)=1$ and $p_r(0) = 0$ for $r \geq 1$), the solution to this equation is another exponential

$$F(z,t) = e^{(\lambda z - \lambda)t} = e^{-\lambda t} \times e^{-\lambda tz}$$

(6.10)

Expanding the second term of Equation 6.10, using the Taylor series expansion, gives

$$F(z,t) = e^{-\lambda t} \sum_r \frac{(\lambda tz)^r}{r!}$$

(6.11)

In Chapter 2, it was shown that the probability of having r events can be derived from the generating function by taking the rth term at $z=1$. Thus the probability of having r arrivals is

$$p_r(t) = \frac{(\lambda t)^r}{r!} e^{-\lambda t}$$

(6.12)

which is a Poisson distribution with an average arrival rate of λt. Note that the mean and variance of this function are equal. Thus the ratio of mean to variance can be used as a measure of randomness, as the closer it is to 1, the more purely random events become.

6.3.3 Holding times

A useful model for random holding times is a uniform probability density, μ, of departure. This implies a negative-exponential distribution. The probability density of the distribution is

$$p(x) = \mu e^{-\mu x}$$

(6.13)

which has mean

$$E(x) = \mu^{-1}$$

(6.14)

and variance

$$var(x) = \mu^{-2}$$

(6.15)

It is useful to note that here the variance is equal to the square of the mean. More general models can therefore use this ratio as a second parameter.

6.4   Traffic process analysis

Analysis of the traffic process in an exchange, say, consists of deriving output statistics in the form of the system occupancy and performance from the measured or postulated input statistics in the form of arrivals or holding times. To do this it is necessary to make the basic assumptions of independence and equilibrium. The independence of the sources is easily understood, but the assumption of equilibrium needs some qualification. Statistical equilibrium means that although the number of occupied devices may fluctuate randomly, the probability of finding a specific number occupied is constant. However, the mean number of devices occupied will vary throughout the day, as illustrated in Figure 6.4, from which it can be seen that statistical equilibrium holds for only relatively short periods. In practice, an uninterrupted period of 60 minutes during which the traffic is at a maximum is known as the busy hour and it is assumed that statistical equilibrium holds during this time. The mean traffic taken over a set of busy hours is referred to as the reference load. Clearly, if the exchange is designed to meet its performance requirements at reference load, it will easily exceed those requirements at lower levels of traffic.

Shorter peaks in traffic may be caused by unusual circumstances such as a sudden fluctuation in stock exchange prices or following a complete breakdown of an exchange when large numbers of subscribers try to re-connect at the same time. However it would be uneconomic to cater for such eventualities except in very special circumstances.

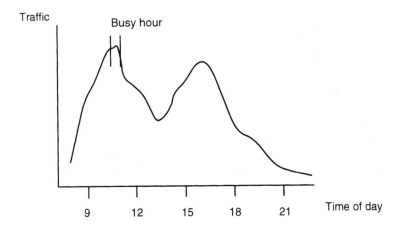

Figure 6.4 Typical diurnal variations in traffic flow

6.5 Traffic problems with simple Markov model

The use of a simple Markov model to analyse the behaviour of a queue was described in Chapter 4. We can now very easily extend the principle to cover circuit switched exchanges. A state, designated by a single number, is used to represent the number of circuits (trunks) in use or the number of calls waiting for service or both. It is assumed that the probability densities for arrivals and departures are constants for any one state. The transition probabilities corresponding to the arrivals and departures are known or at least postulated.

The implications from these assumptions are that both the arrival and departure processes are Poisson, although the rates may be state-dependent. The service time is exponential. The performance of the system in terms of loss and delay may then be determined in relation to the calculated state probabilities.

6.5.1 Infinite source and server

A Markov chain for a hypothetical model representing the case of an exchange having an infinite number of lines and serving an infinite number of subscribers is shown in Figure 6.5. The case of the infinite number of lines is, of course, absurd, but it forms a useful way of introducing the use of Markov chain models. The assumption of an infinite number of subscribers is often a good approximation to the practical condition, in that the probability of a call arriving is independent of the number of calls already connected.

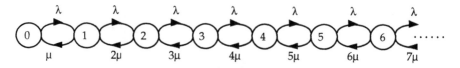

Figure 6.5   Markov chain for infinite source and server

The state number represents the number of calls in the system, the zero state being reserved for the condition in which no calls are being carried. The arrival rate, λ, is constant, since the number of sources is assumed to be infinite. The departure rate, μ, varies with the state, as shown, since the probability of one call clearing is proportional to the number of calls in progress. As shown earlier, at equilibrium, the transition ratio coefficient at state i is defined as

$$D_i = \frac{\text{multiple of upwards transitions}}{\text{multiple of downwards transitions}}$$

which for the above Markov chain becomes

$$D_i = \frac{\lambda^i}{\mu . 2\mu . 3\mu \ldots\ldots i\mu} = \frac{(\lambda / \mu)^i}{i!}$$

or

$$D_i = \frac{a^i}{i!}$$

$$(6.16)$$

where the traffic intensity, $a = \lambda / \mu$.

It was also shown that the state probability distribution is given by

$$p_r = \frac{D_r}{\sum\limits_{i=0}^{\infty} D_i} = \frac{\dfrac{a^r}{r!}}{\sum\limits_{i=0}^{\infty} \dfrac{a^i}{i!}}$$

$$= \frac{a^r}{r!} e^{-a}$$

$$(6.17)$$

which is a Poisson distribution with a mean equal to the traffic intensity, a.

On the assumptions given above, this is the occupancy distribution for negative exponential holding times.

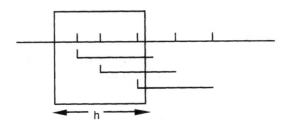

Figure 6.6 Arrivals with constant holding time

If we consider the special case of a constant holding time (see Figure 6.6), then the occupancy is the number of arrivals in the holding time (h). It will be seen that with stationary Poisson arrivals, the occupancy distribution is the same as for negative exponential holding times. In the more general case it may be shown that the occupancy distribution is independent of the holding time.

6.5.2 *Infinite source, finite server – the Erlang loss formula*

Most telephone switching systems are based on the principle that if a call is offered to a switch or a group of trunks and none of the group is free, then the call is cleared and the subscriber must try again. The Markov model of such a system is similar to that for the infinite source, infinite server described above except that the number of states is limited to the number of servers, N (see Figure 6.7).

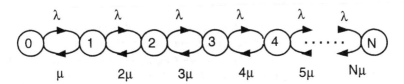

Figure 6.7 Markov chain for infinite source, finite server

In this case

$$D_i = \frac{(\lambda / \mu)^i}{i!} = \frac{a^i}{i!}$$

(6.18)

and

$$p_r = \frac{a^r / r!}{\displaystyle\sum_{i=0}^{N} a^i / i!}$$

(6.19)

A call may arrive at any time. Thus a call arriving when the model is in state N will find all the trunks occupied and that call will be lost. The blocking probability is therefore

$$B(N, a) = p_N = \frac{a^N / N!}{\displaystyle\sum_{i=0}^{N} a^i / i!}$$

(6.20)

This is the well-known Erlang loss or Erlang B formula. The expression is well tabulated in many publications. It is useful to note the bounds

$$\frac{a^N}{N!} e^{-a} < B(N, a) < \frac{a^N}{N!}$$

(6.21)

As with the infinite source and server, the occupancy distribution is invariant with holding time distribution.

6.5.3   Grade of service

In telephony, the loss probability during the busy hour is known as the grade of service (GoS) and is an important factor in specifying the performance of a switching system. Figure 6.8 shows the blocking probability for full-availability groups against the loading per trunk calculated from Equation 6.20. It will be seen that large groups of trunks are more efficient in carrying traffic (i.e. they have a higher value of a/N for the same blocking probability), as might be expected. However, they are also more sensitive to overload (i.e. the slope of the graph is steeper for larger groups).

As explained earlier, the mean busy hour traffic may fluctuate, so that overload conditions should be considered. For example, a typical specification might include

  – blocking probability at reference load (say 0.01)

- blocking increased by < 2 times at 10% overload

- blocking increased by < 5 times at 20% overload

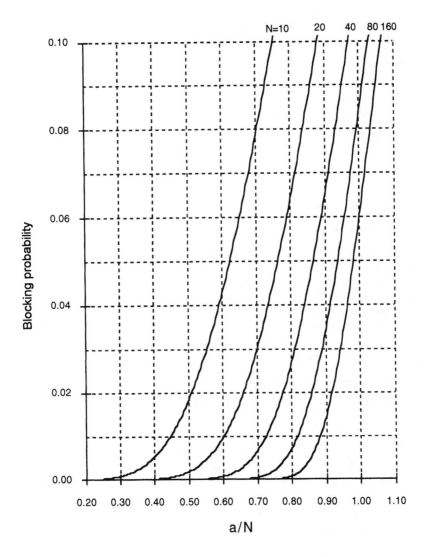

Figure 6.8 Blocking in full availability groups

Table 6.1 shows how differently sized systems would perform under overload conditions if the normal traffic in each case is the maximum that could be carried on that number of trunks with 1% blocking probability.

Table 6.1   Effects of overload

No of trunks (N)	Normal traffic (Erlang)	Blocking 10% overload	Blocking 20 % overload
10	4.46	0.017	0.026
20	12.0	0.020	0.036
40	29.0	0.026	0.052
80	65.4	0.033	0.074

A system based on ten or twenty trunks, if designed for the reference load, would meet the criteria for 10% and 20% overload. However, a 40-trunk system would fail to meet the 10% overload condition and also the 20% condition, although in the latter case it might just be acceptable if the high overload were expected to occur only very rarely. The design traffic would have to be reduced to 28.2 Erlangs to meet the specification. In the case of the 80-trunk system the traffic would have to be reduced to 62.4 Erlangs to meet the 10% overload condition and to 62.3 Erlangs to meet the 20% condition. Clearly then, as the number of trunks in a group increases, it will be the overload specification that will limit the amount of traffic carried.

6.5.4   Finite source and server: the Engset and binomial distributions

In all the cases discussed so far, it has been assumed that the number of sources was infinite, so that the probability of a call arising did not change with the number of calls already in progress. This is an acceptable assumption for a large number of sources, but falls down if the number of sources is less than 100, say. For example, if we have ten sources and five of them are already busy, the probability of a new call is one half of what it would be if no calls were in progress. The Markov chain for this case is shown in Figure 6.9. The number of sources is S, the number of trunks is N, $(N<S)$, and the arrival rate per source is ρ. The probability of a call being initiated when all the trunks are free is $S\rho$, but when the first trunk is engaged, the call probability falls to $(S-1)\rho$ and so on.

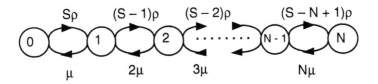

Figure 6.9 A Markov chain of finite source and server

Here again we find

$$D_i = \frac{S\ (S-1)(S-2)\ldots\ldots(S-i+1)\rho^i}{\mu\ .\ 2\mu\ .\ 3\mu\ \ldots\ldots\ i\mu}$$

or

$$D_i = (\frac{\rho}{\mu})^i \begin{bmatrix} S \\ i \end{bmatrix}$$

(6.22)

and with $h=1/\mu$,

$$D_i = (\rho h)^i \begin{bmatrix} S \\ i \end{bmatrix}$$

and the blocking probability,

$$p_r = \frac{(\rho h)^i \begin{bmatrix} S \\ i \end{bmatrix}}{\sum_{i=0}^{N} (\rho h)^i \begin{bmatrix} S \\ i \end{bmatrix}}$$

(6.23)

This is the Engset distribution.

A special case arises if the number of inlets to a switch equals the number of outlets, i.e. S=N. Then

$$P_r = \frac{(\rho h)^r \begin{bmatrix} S \\ r \end{bmatrix}}{(1+\rho h)^S} = \begin{bmatrix} S \\ r \end{bmatrix} p^r (1-p)^{S-r}$$

with $p = \rho h / (1+\rho h)$

(6.24)

This is an example of the binomial distribution discussed in Chapter 2; p is the probability of an individual trunk being busy.

6.6   Delay systems

Although most automatic telephone systems are based on the principle that calls that cannot be connected immediately are lost, it is possible to consider systems in which such calls are delayed until a free trunk becomes available. In fact, such conditions occurred in the early days of manual exchanges when over a short period of time, calls might appear at the manual operator's switchboard at a faster rate than the operator could deal with them. He or she would attempt to deal with them in turn until the temporary overload subsided. More recently, in some enquiry systems the calls are queued automatically until an operator is available to deal with them. Simple queuing systems have already been covered in chapter 4.

6.6.1   Queue with infinite buffer

Let us assume a single queue with infinite source as a principal example, shown in Figure 6.10.

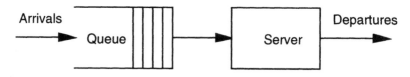

Figure 6.10 A single server with infinite arrivals

We also assume an unbounded queue space, whose chain model is shown in Figure 6.11.

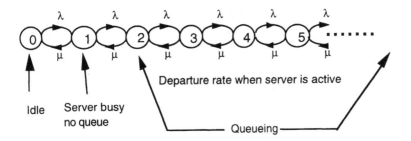

Figure 6.11 Single server with infinite buffer

The general transition coefficients for such a queue are

$$D_i = (\lambda / \mu)^i = a^i$$

(6.25)

and the state probability is

$$p_r = \frac{a^r}{\sum\limits_{i=0}^{\infty} a^i} = a^r (1-a)$$

(6.26)

The system is in equilibrium only if $a<1$ or equivalently $\lambda<\mu$.

We are interested in calculating the mean delay of the system. All delays will be in units of mean holding time $h=1/\mu$. For convenience, we normalise the system to $\mu=1$: then the unit of delay will be unit holding time. The probability of having any delay in the system is the complement of not having a delay (i.e. zero delay), thus, the delay probability, p_D, is

$$p_D = 1 - p_0$$

(6.27)

with p_0 being the probability of zero delay. From Equation 6.26, the probability of zero delay, i.e. $r = 0$, is $p_0 = 1 - a$. Hence the delay probability

$$p_D = a$$

Assume that the queue operates on first-in first-out basis. Thus an arrival in state i waits for i users, so it has a mean queuing time i. Then the mean delay is the average of all delayed calls, which is (using conditional probabilities)

$$E(w_D) = \frac{\sum\limits_{i=1}^{\infty} i\, p_i}{\sum\limits_{i=1}^{\infty} p_i}$$

(6.28)

Substituting for the state probability from Equation 6.26

$$E(w_D) = \frac{\sum\limits_{i=1}^{\infty} i\, a^i (1-a)}{\sum\limits_{i=1}^{\infty} a^i (1-a)}$$

which can be simplified to

$$E(w_D) = \frac{1}{1-a}$$

(6.29)

This is the queuing time spent by those who are delayed. For an offered traffic of a, then average queuing time over all arrivals is

$$E(w_Q) = \sum\limits_{i=0}^{\infty} i p_i = \frac{a}{1-a}$$

The total time spent in the system has mean

$$E(w) = E(w_Q) + 1$$

$$= \frac{a}{1-a} + 1 = \frac{1}{1-a}$$

(6.30)

That is, the mean total time spent (delay) in the system is the sum of the queuing delay plus the unit holding time. In Figure 6.12 the mean total and queuing delays are plotted for various values of traffic load a.

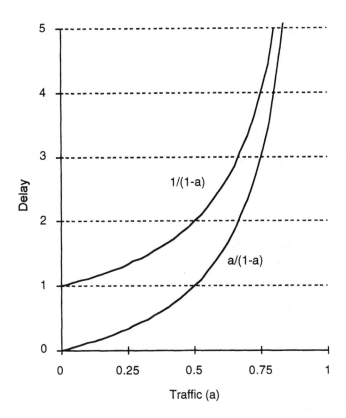

Figure 6.12 Total and queuing delay for infinite buffer

6.6.2 Waiting time of delayed calls

Since the mean delay is constant, it is expected that the waiting time of delayed calls will have a negative-exponential distribution (see Chapter 2, Section 2.4.5). Here again we verify that it is the case.

Let us assume that in the time interval (t, t+dt), there are r arrivals, with the rth arrival occurring at time t+dt, as shown in Figure 6.13.

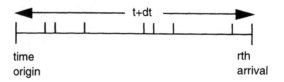

Figure 6.13   An interval with r arrivals

Thus there are r−1 arrivals in (0, t) and one arrival in (t, t+dt). The probability of one arrival in (t, t+dt) is 1dt, for normalised arrival rate 1. Further, we have shown that the probability of the number of arrivals in a given time interval follows a Poisson distribution. Hence the probability density of the sum of r negative exponential intervals is

$$g_r(t)dt = prob[(r-1)\ events\ in\ (0, t)] \times prob[event\ in\ (t, t+dt)]$$

$$= \frac{t^{r-1}}{(r-1)!} e^{-t} \times 1dt$$

(6.31)

Now, an arrival in state i has to wait for i users, and so has a waiting time density $g_i(t)$. Averaging over all delayed calls, the waiting time density is

$$f(t) = \frac{\sum_{i=1}^{\infty} p_i g_i(t)}{\sum_{i=1}^{\infty} p_i}$$

(6.32)

using conditional probabilities, where p_i is the unconditional probability of being in state i. Substituting for p_i and $g_i(t)$ from Equations 6.26 and 6.31, yields

$$f(t) = \frac{\sum_{i=1}^{\infty} a^i (1-a) \frac{t^{i-1}}{(i-1)!} e^{-t}}{\sum_{i=1}^{\infty} a^i (1-a)}$$

(6.33)

This equation may then be simplified to

$$f(t) = (1-a) e^{-(1-a)t}$$

(6.34)

Note that this is a negative-exponential function with mean = $1/(1-a)$, as was shown previously.

6.6.3 Queue with finite buffer; blocking and delay

If in the previous example the buffer size was limited to M positions, then the excess traffic applied to the queue would have to be rejected, or access to the queue blocked. The blocking in this case is shown in Figure 6.14 by a simple rejection switch.

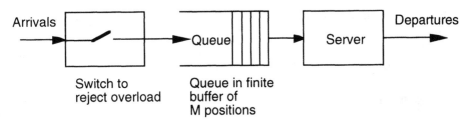

Arrivals — Switch to reject overload — Queue — Queue in finite buffer of M positions — Server — Departures

Figure 6.14 Queue with finite buffer size

The queuing model for this figure is the M+2 state Markov chain, shown in Figure 6.15.

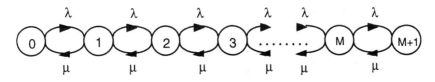

Figure 6.15 Queueing model of finite buffer with a blocking switch

As shown in Section 4.8.5 of Chapter 4, the state probability for this model is

$$p_r = \frac{a^r (1-a)}{1-a^{M+2}}$$

(6.35)

where state zero ($r = 0$) represents immediate service with no delay

$$p_s = p_0 = \frac{1-a}{1-a^{M+2}}$$

(6.36)

The last state, $r = M+1$, represents the blocking state, with a blocking probability

$$p_B = p_{M+1} = \frac{a^{M+1}(1-a)}{1-a^{M+2}}$$

(6.37)

For $a < 1$, the blocking probability p_B approaches zero as the buffer size becomes larger, i.e.

$$p_B \rightarrow 0, \quad \text{as } M \rightarrow \infty$$

and for $a > 1$

$$p_B \rightarrow 1-1/a, \quad \text{as } M \rightarrow \infty$$

Thus $1-1/a$ is the upper bound for the blocking probability. Figure 6.16 shows the blocking probability for various buffer sizes and traffic intensity a.

The delay probability can be calculated by summing the probability of the states between 0 and $M+1$, i. e.

$$p_D = \sum_{i=1}^{M} p_i$$

$$p_D = \sum_{i=1}^{M} \frac{a^i(1-a)}{1-a^{M+2}}$$

(6.38)

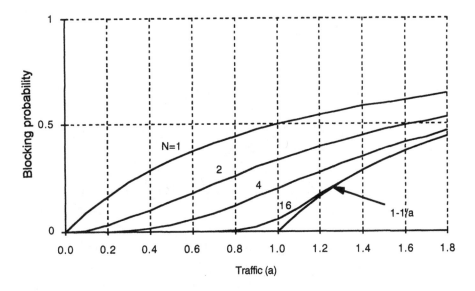

Figure 6.16 Blocking probability

Note that here the maximum queue state is M, not M+1. The mean queuing time (considering all states but the blocking state M+1) in unit holding times (i.e. $\mu=1$) is

$$E(W_Q) = \sum_{i=0}^{M} i p_i \Big/ \sum_{i=0}^{M} p_i = \sum_{i=0}^{M} \frac{ia^i(1-a)}{1-a^{M+2}} \Big/ \sum_{i=0}^{M} \frac{a^i(1-a)}{1-a^{M+2}}$$

$$= \sum_{i=0}^{M} i a^i \Big/ \sum_{i=0}^{M} a^i$$

(6.39)

Since

$$\sum_{i=0}^{M} i a^i = a \frac{d}{da} \sum_{i=0}^{M} a^i$$

$$= \frac{a}{(1-a)^2} [1 - (M+1) a^M + Ma^{M+1}]$$

(6.40)

and

$$\sum_{i=0}^{M} a^i = \frac{1-a^{M+1}}{1-a}$$

then the mean queuing delay is

$$E(w_Q) = \frac{1-a}{1-a^{M+1}} \cdot \frac{a}{(1-a)^2} [1 - (M+1) a^M + M a^{M+1}]$$

$$= \frac{a}{(1-a)} \cdot \frac{[1-(M+1) a^M + M a^{M+1}]}{1-a^{M+1}}$$

$$= \frac{a}{1-a} - \frac{(M+1)a^{M+1}}{1-a^{M+1}}$$

$$(6.41)$$

Now for $a < 1$, and large buffer sizes, $M \to \infty$, the mean queuing delay becomes

$$E(w_Q) = \frac{a}{1-a}$$

$$(6.42)$$

And if the traffic intensity becomes very large, $a \to \infty$, then the mean queuing delay approaches the full buffer size, $E(w_Q) \to M$. Figure 6.17 shows the variation of the mean queuing delay for various buffer sizes and traffic intensity.

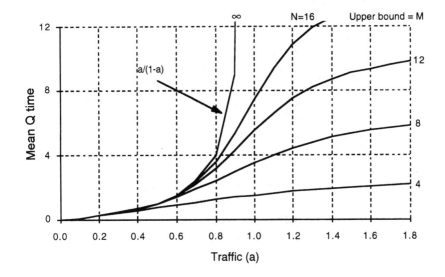

Figure 6.17 Mean queuing delay

Example

In a mobile telephone net serving a fleet of ambulances, calls are generated at random at an average rate of 4 calls per minute, and each call lasts an average of 20 seconds. Calls that cannot be dealt with immediately are assumed to be queued. Normally the system is controlled by 2 dispatchers, but while 4 or more calls are in the system (i.e. calls in progress plus calls queued) the supervisor takes over as a third dispatcher position.

(a) Draw a Markov diagram to represent the possible states of the system, showing the transition probabilities.

(b) Find the probability that a new call will have to wait for service.

(c) Calculate the percentage of the time the supervisor is engaged in dealing with calls.

(a) If time is measured in seconds, the probability of a call arriving in time dt is

$$= (4/60) \times dt = 0.067\ dt$$

Hence the arrival rate is fixed at $\lambda = 0.067$.

The probability of a dispatcher clearing a call in dt is

$$= (1/20) \times dt = 0.05\ dt$$

Therefore if the state of the queue is such that one dispatcher is currently dealing with a call, then the service rate μ is

$$\mu = 0.05$$

When both dispatchers are dealing with calls, then the service rate is

$$\mu = 0.05 \times 2 = 0.1$$

Finally, when there are 4 or more calls in the system, the supervisor has to take calls, and so the service rate becomes

$$\mu = 0.05 \times 3 = 0.15$$

Hence the Markov diagram representing the states of the system is

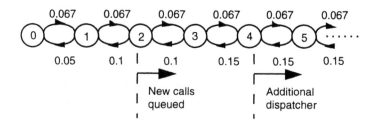

with the transition probabilities as shown.

(b) To calculate the waiting time, the state probabilities have to be calculated. This is done in the equilibrium condition where

$$p_i \times b_{i \to i+1} = p_{i+1} \times d_{i+1 \to i}$$

Hence

$$p_1 = \frac{0.067}{0.05} p_0 = 1.3333 p_0$$

$$p_2 = \frac{0.067}{0.05} p_1 = 0.8889 p_0$$

$$p_3 = \frac{0.067}{0.1} p_2 = 0.5926 p_0$$

$$p_4 = \frac{0.067}{0.15} p_3 = 0.2634 p_0$$

and so on

Summing the states above 4 to infinity (assuming an infinite buffer, as shown in the figure)

$$p_{4\to\infty} = \frac{0.2634p_0}{1 - \frac{0.67}{0.15}} = 0.4741p_0$$

Since the sum of the state probabilities is unity,

$$p_0 + p_1 + p_2 + \ldots\ldots\ldots + p_\infty = 4.5523p_0 = 1$$

then

$$p_0 = \frac{1}{4.5523} = 0.2197$$

and

$$p_1 = 1.3333p_0 = 0.2929$$

A call will have to wait in the queue if it arrives when the model is in state 2 or higher. Hence the probability of a call having to wait

$$p(w) = 1 - p_0 - p_1 = 0.4874$$

(c) The supervisor will man the operator position when the system is in state 4 or higher with probability

$$p_{4\to\infty} = 0.4741p_0 = 0.1041$$

that is, about 10% of the time.

6.7 Exchange trunking

In circuit switched networks a physical path or circuit must connect an input and its designated output for the duration of the call. In analogue switching, such a connection is established by space switches. The switch connects and disconnects physical contacts using a matrix of crosspoints. The connection is permanent for the duration of the call. For digital switches, the connection is only made during a time slot, and in the next slot the digital space switch may be used to connect other channels. However, since the positions of the incoming and outgoing time slots (channels) are fixed, then the connect/disconnect function of the switch is repeated cyclically at frame rate (e.g. every 125 μs).

However, digital space switches cannot connect any two channels on the same multiplexed line. In this case 'time-slot-interchange', or for short 'time switches' are used. In a time switch, the incoming slots of a multiplexed line are written into a memory, in the order they arrive, but are read out in the order desired for the interchange (connection). Obviously, though, time switches cannot connect slots from two different multiplex routes, and so must be used in combination with space switches. For all but the smallest exchanges it is clearly uneconomic to provide a single block of switches such that any inlet could be connected to any outlet. A more usual approach is to employ a number of switch blocks in tandem. A typical arrangement is shown in Figure 6.18.

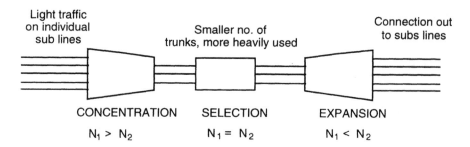

Figure 6.18   Three switch blocks in tandem

Apart from the economy in the number of crosspoints a multi-stage switching arrangement also has the advantage that diverse technologies can be combined. For example in analogue systems, the switching blocks of Figure 6.18 are all space switches with permanent connection. In digital switches, the requirement of the switch to be able to connect any channel of one incoming inlet to any channel of any outlet is met by using a combination of space and time switches. In Figure 6.18, the first column of switch matrices might be time switches, the second space switches and the third time switches, to give a common arrangement that is sometimes abbreviated T-S-T. An example of a 3-stage switch with 9 inlets and 9 outlets is shown in Figure 6.19.

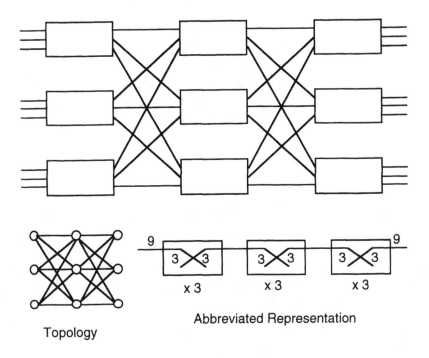

Figure 6.19 Representation of multi-stage switching

At each stage, each switching block is a 3x3 switching matrix, providing at least one path for every inlet. The full representation of the arrangement shown in Figure 6.19 may be abbreviated as shown. Sometimes it is convenient to represent the internal linking between the stages with a connection topology, as is also shown in the figure.

6.8 Multi-stage networks

6.8.1 Non-blocking networks

Consider the multi-stage network shown in Figure 6.20 in which a given inlet in an A switch array has to be connected to a given outlet in the C switch array. Suppose (c – 1) inlets in the same A switch are busy and that (c – 1) outlets in the desired C switch are also busy. In the worst case, the busy inlets and the busy outlets use different arrays in the B switch. Then for a free path to be available, the call must be routed through an unused B switch array. Since each of the b trunks from an A switch is connected to a different B switch, for a B switch to be available for the new call,

$$b \geq (c-1) + (c-1) + 1 = 2c - 1$$

This is known as the non-blocking or Clos condition. This condition can be easily visualised, using a channel graph, as shown in the figure. If however, it is possible to rearrange the calls in progress to accommodate the new call, the only condition for the new call to be connected is that there is always at least one free B switch, that is

$$b \geq c$$

6.8.2 Multi-stage networks with blocking

If the Clos condition is not met, it is necessary to consider the probability of a new call being blocked because no free path is available between the inlet and the required outlet. The problem is quite difficult because, apart from the combinatorial problems caused by the many potential paths in large exchanges, the dependencies between the blocking probabilities on the different links make exact calculations almost impossible. It is therefore necessary to use approximations. The simplest approach is to assume that the probability of a link being occupied is independent of all other factors. Let this probability be p. Referring to Figure 6.20, the probability of both links from the required A and C switches to a particular B switch being free is

$$(1 - p)^2$$

and therefore the probability of that path being blocked is

$$1 - (1 - p)^2$$

The probability that there is no free path through any of the B switches is

$$p_B = [1 - (1-p)^2]^b$$

$$(6.43)$$

Since we have assumed uniform loading, $p = ac/b$ where a is the level of traffic on each of the inlets (the 'calling rate').

Then

$$p_B = \left[1 - \left(1 - \frac{ac}{b} \right)^2 \right]^b$$

$$(6.44)$$

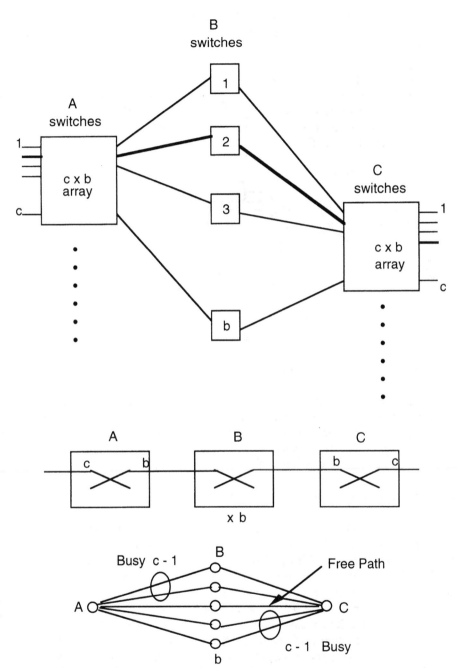

Figure 6.20 A three-stage switch and its equivalent channel graph

This is the approximate blocking probability for any nominated terminal pair if the loading is independent and uniform. The true blocking probability is less than that given by the approximate model, so that Equation 6.43 is useful as an upper bound. Note that if we apply the minimal Clos condition, $b = 2c - 1$, to Equation 6.43, the blocking probability does not fall to zero.

The above analysis is general, and the approximate blocking probability can be applied to any combination of space and time switches. The correspondence between the three stage all-space switching arrangement of Figure 6.20 and a TST digital system is shown in Figure 6.21.

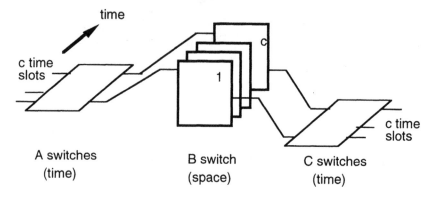

Figure 6.21   A time-space-time switch

Instead of an A switch with c inlets, we now have a single multiplex of c channels. Each A switch becomes a time switch such that the contents of any time slot on the inlet may be shifted in time to any free time slot on the switch outlet. The B switch consists of a single matrix in which the connections may be changed for each time slot. The C switch arrangement is similar to that of the A switch.

6.9   General multi-stage analysis

Early telephone switching systems often operated using a step-by-step process, such that connections were made, as required, in successive stages. Under these conditions, the overall blocking probability can be approximated as the product of the blocking probabilities at each stage. More modern systems, however, use a conditional selection process, such that a free outlet through the entire multi-stage switch must be located before any connections are made.

Although there will be many possible paths through such a multi-stage switch, traffic from a large number of random sources has to be carried, and blocking can still occur if all the paths between a given inlet/outlet pair are occupied. Let the event that all such paths are occupied be Y. The probability P(Y) is difficult to obtain in general, as the large number of stages and transitions between them make it impractical to use a Markov model. However, we may be able to tackle the problem using conditional probabilities. If we fix some variable X, say the number of busy links in a group, the conditional blocking probability $P(Y|X=x)$ may be much easier to calculate [4].

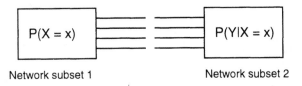

Network subset 1   Network subset 2

Figure 6.22   Conditional probabilities for switching matrices in tandem

For example, with reference to the two network subsets shown in Figure 6.22, suppose the probability that network subset 1 is in state x is equal to $P(X = x)$. Now consider the probability of blocking in network subset 2 given that network subset 1 is in state $X = x$, i.e. $P(Y|X=x)$. Then the overall probability of blocking is

$$P(Y) = \sum_{x} P(X = x) \cdot P(Y|X = x)$$

(6.45)

6.9.1   Two links in tandem

Before we look at a problem where we must use conditional probabilities to obtain a solution, let us first consider a simple example to illustrate the method.

Consider a three-stage switching network, with the group of links between the first and second stages designated the B links, and those between the second and the third the C links. If there are m links in each group, the channel graph would be as shown in Figure 6.23. Let $G(x)$ be the probability that precisely x of the C links are busy, and $H(y)$ the probability that a given set of y B links do not contain a free link, with $H(0)=1$.

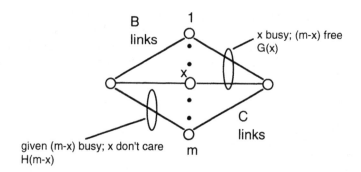

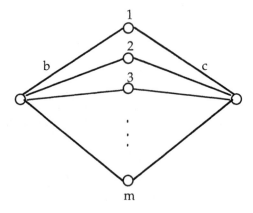

Figure 6.23   Channel graph for two switching stages in tandem

From our model above, taking the state X to be the number of C links, the overall blocking probability is

$$P_B = \sum_{x=0}^{m} G(x) \, H\,(m-x)$$

(6.46)

Now suppose the B and C links have binomial distributions with link occupancies b and c respectively, as shown in Figure 6.24.

Figure 6.24   Link occupancy

The assumption of a binomial distribution is often valid for well smoothed traffic although G and H may have Erlang or Engset forms, according to the specific situation.

Then

$$G(x) = \binom{m}{x} c^x (1-c)^{m-x}$$

$$H(x) = b^x$$

$$(6.47)$$

The blocking probability, p_B

$$
\begin{aligned}
p_B &= \sum_{x=0}^{m} \binom{m}{x} c^x (1-c)^{m-x} b^{m-x} \\
&= (1-c)^m b^m \sum_{x=0}^{m} \binom{m}{x} \left(\frac{c}{(1-c)b} \right)^x \\
&= (b-bc)^m (1 + \frac{c}{b-bc})^m
\end{aligned}
$$

$$(6.48)$$

Hence the blocking probability is

$$p_B = (b + c - bc)^m$$

$$(6.49)$$

However, this result can also be derived in a more straightforward way. Consider the channel graph of Figure 6.24 to have independent link occupancy. Thus any one of the m parallel paths would be free if both its links were free, therefore it would be blocked with probability $1 - (1 - b)(1 - c)$. Further, the full channel graph would be blocked if all its individual paths were blocked and so by multiplication of the independent probabilities

$$p_B = \{1 - (1-b)(1-c)\}^m = (b + c - bc)^m$$

6.9.2   *Connections other than series or parallel*

Having considered an example which can be solved more simply without the use of conditional probabilities, let us now consider a multi-stage switching network that is not connected in a straightforward series or parallel manner. A channel graph corresponding to such an arrangement is shown in Figure 6.25, after Takagi [7].

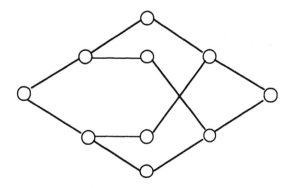

Fig. 6.25   A mixed parallel and serial channel graph arrangement

We will assume each link is independently busy with probability p, and free with probability q = 1 – p. The blocking probability of this channel graph cannot be found from series/parallel calculations. Although other methods exist, such as the use of network paths or cuts, consider instead enumerating only the configurations of blocked and free links in the two groups of links adjacent to the terminal points. The channel graph conditional on each configuration is a simple series/parallel graph whose conditional blocking probability can be calculated in terms of b = 1 – q^2, i.e. the blocking probability of two tandem links. The performance may be analysed in terms of the earlier conditional probability model, but in this case taking X to be the configuration of blocked and free links (see Table 6.2).

From the table, blocking probability can be calculated, knowing p_x and b_x for each value of x, that is

$$p_B = \sum_{x=1}^{6} p_x b_x$$

$$= q^4 b^4 + 4q^3 pb^2 + 2q^2 p^2 (2b+1) + 4qp^3 + p^4$$

$$\text{(6.50)}$$

Table 6.2 Switching configurations

x	configuration	prob. of occurence (p_x)	conditional channel graph	blocking prob (b_x)
1		q^4		b^4
2		$4q^3p$		b^2
3		$4q^2p^2$		b
4		$2q^2p^2$		1
5		$4qp^3$		1
6		p^4		1

indicate a busy link

indicate a free link

6.10 Multi-stage access to trunk groups

In some cases, we are concerned with the probability of blocking not for the connection between a specific inlet and a specific outlet, but rather between the specific inlet and any one of a number of outlets. This condition occurs for a connection between a calling subscriber and a particular trunk or junction route to another exchange. There will normally be several trunks offering a connection to the next exchange and any free trunk may be selected. A typical arrangement is shown in Figure 6.26, where **200** inlets need to be routed into routes k, l and m.

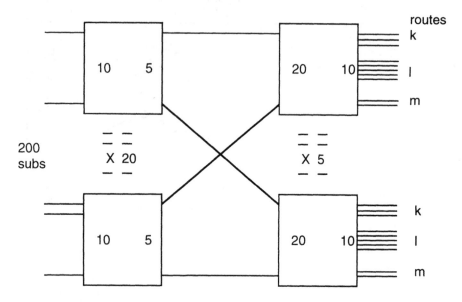

Figure 6.26   Two-stage network for connection to trunk groups

A call may originate from any one of the **200** subscribers and a connection may be required to one of three trunk routes, k, l and m. Suppose route k has **15** circuits, route l has **25** circuits and route m has **10** circuits. These circuits are spread as evenly as possible over the five switch matrices as shown. This approach gives lower blocking probability than the alternative of concentrating all the circuits on a route on a smaller number of the matrices.

The channel graph for a connection from any inlet to destination k is shown in Figure 6.27.

Then assuming independence and equiprobable loading of the links and trunks, the probability that a stage 1 to stage 2 link is free is $(1-p)$ and the probability that a stage 2 to destination route k trunk is free is $(1-q)^3$. Then the approximate blocking probability for a connection to route k is given by;

$$p_B = \{1-(1-p)(1-q^3)\}^5$$

One point to note is that, due to concentration of traffic in the second stage q=2p, and further, due to concentration in the first stage, that p is twice the traffic offered by an individual subscriber.

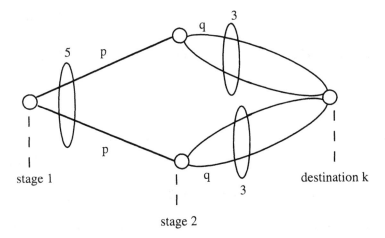

Figure 6.27 A channel graph for two stage connection

6.11 Overflow traffic

6.11.1 Characteristics of overflow traffic

In step-by-step exchanges, traffic that could not be carried by a particular grading group was applied to the next group. As explained earlier, this is not important nowadays since it is usually possible to preselect a path through the exchange. However the principle is still important when traffic that cannot be carried on the direct route to another exchange is diverted to an indirect route via a third exchange, a technique known as alternative routing.

One straightforward way of performing alternative routing is simply to have a first choice group of circuits and then a second group to handle (possibly pooled) overflow traffic [4]. Clearly the traffic on the overflow group will not be Poisson, consisting as it does of the bursts of otherwise blocked traffic from the first group. We will thus model the overflow traffic as short intervals of random length (negative exponential) during which the traffic overflows. Further, we will assume that the onset of these bursts is itself at random (a Poisson process), as shown in Figure 6.28.

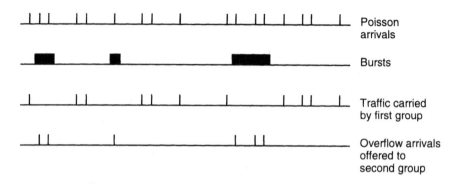

Figure 6.28   Random bursts of overflow traffic

To find the number of calls in a burst, consider a burst of duration t. Then the conditional generating function (Poisson) is

$$F(z, t) = e^{(\lambda z - \lambda)t}$$

(6.51)

But the duration has a negative exponential distribution

$$f(t) = \beta e^{-\beta}$$

(6.52)

with expectation

$$E(t) = \frac{1}{\beta}$$

(6.53)

Hence the unconditional generating function is

$$\int_0^\infty e^{(\lambda z - \lambda)t} \beta e^{-\beta t} dt = \frac{\beta}{\beta + \lambda - \lambda z}$$

$$= \frac{1 - \rho}{1 - \rho z}$$

(6.54)

where $\rho = \dfrac{\lambda}{\lambda + \beta}$

This describes a geometric distribution with

$$P_r = (1-\rho)\rho^{\,r} \tag{6.55}$$

where r is the number of call arrivals in a burst.

Overflow traffic is the summed effect of all such bursts. Its generating function is approximately

$$\left[\dfrac{1-\rho}{1-\rho z}\right]^{\gamma\,h} \tag{6.56}$$

where γ is the mean burst rate and h is the mean call holding time.

This is a negative binomial distribution with

$$\dfrac{\text{variance}}{\text{mean}} = \dfrac{\gamma\, h\rho\ (1-\rho)}{(1-\rho)^2\ \gamma\, h\rho} = \dfrac{1}{1-\rho}$$

$$= 1 + \dfrac{\lambda}{\beta} > 1 \tag{6.57}$$

where λ is the call arrival rate and β is the rate at which bursts complete.

The variance/mean ratio is thus greater than 1, unlike both pure Poisson, which has a ratio of exactly 1, and smoothed traffic which has a ratio less than 1. (Smoothed traffic is simply what remains after some of the peaks have been removed from a traffic stream by earlier blocking.) Consequently, because of this peakiness of overflow traffic, a larger number of circuits is required to carry the same amount of traffic, at an acceptable blocking probability, than would be required for Poisson traffic.

The above theory depends on the assumptions stated and therefore the results can only be regarded as approximate. However, more precise theories show that a two parameter distribution can match overflow traffic very well.

6.11.2   *Dimensioning of overflow groups*

In some cases, the overflow traffic from more than one group can overflow on to a single set of trunks. The following treatment is based on a two-parameter distribution for the overflow traffic. The mean and variance of the traffic is matched to either a negative-binomial or equivalent random source.

Let us assume there are k high usage routes, each carrying a traffic load of a_i Erlang into n_i trunks. The overflow traffic is then carried by a single route of n trunks, as shown in Figure 6.29.

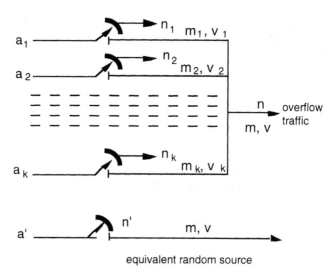

Figure 6.29   Dimension of overflow traffic

The mean traffic lost from each incoming route can be calculated using the Erlang loss formula. For the ith incoming route the mean loss traffic is

$$m_i = a_i B(n_i, a_i) \qquad \text{Erlang}$$

(6.58)

It can be shown [1] that the variance of the overflow traffic from such a route is given by

$$v_i \cong m_i \left[\frac{a_i}{n_i - a_i + m_i + 1} - m_i + 1 \right]$$

(6.59)

The mean and variance of the pooled traffic on the overflow route is

$$m = \sum_{i=1}^{k} m_i$$

$$v = \sum_{i=1}^{k} v_i$$

The equivalent random traffic on the overflow route is [8]

$$a' \cong v + \frac{3v}{m}\left[\frac{v}{m} - 1\right]$$

(6.60)

offered to n' circuits, where n' is given by

$$n' \cong \frac{a'}{1 - (m + \frac{v}{m})^{-1}} - m - 1$$

(6.61)

Hence the equivalent random source to the overflow route offers traffic of a' Erlang into n' circuits, as shown in Figure 6.29.

The procedure is then to calculate n such that the lost traffic $a' B(n, a')$ meets the stated requirements. The minimum size of the overflow group is then $n - n'$. Under these conditions, it is found that the first choice circuits can be used very efficiently and loadings of about 80% are common. The overflow circuits have a much lower efficiency, typically 40% due to the peaky nature of the traffic.

6.11.3 Alternative routing strategies

So far, only a simple overflow situation has been considered. In a network of exchanges, more complex overflow strategies may be used. Some examples are given in Figure 6.30.

Analysis of the simple overflow, shown in Figure 6.30.a, is straightforward. In this case the overflow traffic from the high usage route 1, is simply carried into route 2. The required number of circuits on this route to cope with the overflow traffic can be calculated as described above. Analysis of traffic with mutual overflow, as shown in Figure 6.30.b, is more complex than this simple case. Here, a high usage route connecting any two nodes, is itself a tandem route for other two nodes. With more complex mesh and hierarchical connections, shown in Figures 6.30.c and 6.30.d the analysis is even more complex. In the complete mesh of Figure 6.30.c, the overflow traffic on one link can be diverted into $\frac{1}{2}N(N-1)-1$ other links, which themselves are high usage links for their own main routes. A similar situation exists in Figure 6.30.d, where in addition to complete mesh between the nodes, the hierarchical structure of the connection makes the analysis even more difficult.

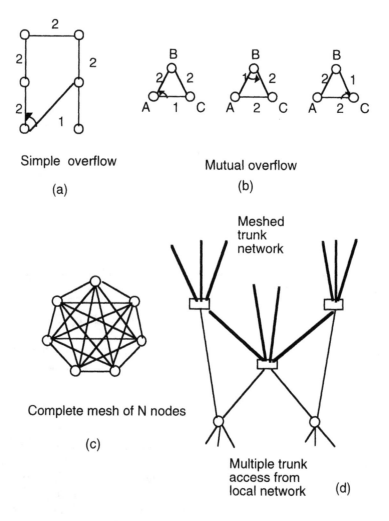

Simple overflow

(a)

Mutual overflow

(b)

Meshed trunk network

Complete mesh of N nodes

(c)

Multiple trunk access from local network   (d)

Figure 6.30 Overflow strategies

6.11.4 Blocking with mutual overflow

Consider a group of exchanges with mutual overflow. Communication between any two exchanges is possible both by a direct link and also by r 2-link paths as shown in Figure 6.31.

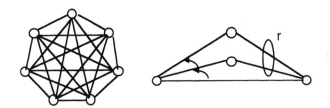

Fig. 6.31 Direct and r 2-link paths

Let a_1 be the traffic carried on each link and B_1 the blocking probability of a link with traffic a_1. If the blocking probability on a link, B_1, is known, the combined blocking probability of the r overflow paths is

$$B_2 = [1 - (1-B_1)^2]^r$$

$$= B_1^r (2 - B_1)^r$$

(6.62)

Then the total blocking of the traffic stream between two exchanges is

$$B = B_1 \times B_2$$

$$= B_1^{r+1} (2 - B_1)^r$$

(6.63)

The mean path length can be found by considering the traffic flows between two exchanges.

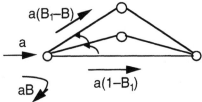

Figure 6.32 Calculation of mean path length

The traffic on the r overflow paths uses two links, whereas that on the direct link uses only one.

$$L = \frac{1 \times a(1-B_1) + 2 \times a(B_1 - B)}{a(1-B)}$$

$$= \frac{(1-B_1) + 2(B_1 - B)}{1-B}$$

$$= \frac{1 + B_1 - 2B}{1-B}$$

$$(6.64)$$

The general behaviour of the blocking probabilities and the mean path length as a function of traffic loading on a particular link are shown in Figure 6.33. As the loading a_1 is increased, the blocking probability for the direct link increases steadily but the total blocking increases slowly at first, but then rises rapidly as all the alternative routes become blocked. The mean path length increases steadily with traffic but when the loading is increased beyond a critical point, the path length decreases because of the high blocking probability.

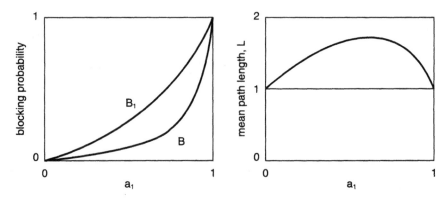

Figure 6.33   General behaviour of systems with mutual overflow

The traffic carried per stream is

$$a' = \frac{a_1}{L}$$

$$(6.65)$$

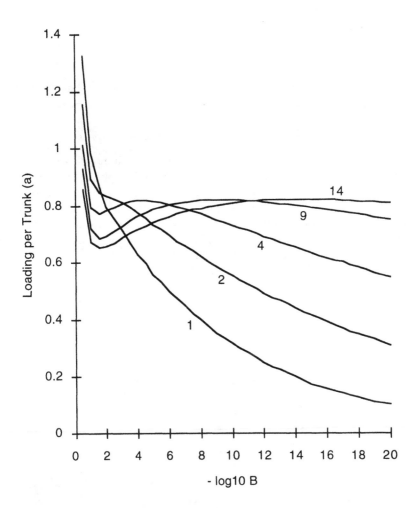

Fig. 6.34   Example of equilibrium characteristic of overflow system

The traffic offered per stream is

$$a = \frac{a_1}{L(1 - B)}$$

(6.66)

Substituting for the mean path length

$$a = \frac{a_1}{1 + B_1 - 2B}$$

(6.67)

Figure 6.34 shows the traffic offered for a group of 20 trunks per link, for various numbers of alternative paths (r+1).

6.11.5   *Dynamic behaviour of mutual overflow systems*

Mutual overflow systems exhibit unstable behaviour at high traffic loadings. From the equation of traffic offered per stream (Equation 6.67), a plot of the equilibrium characteristic of total blocking, B, against the traffic offered may have a characteristic such as shown in Figure 6.35.a. Such multivalued equilibrium characteristics normally imply hysteresis in dynamic behaviour such that the proportion of direct paths used for the traffic can remain in a quasi-stable state for 100 or more call holding times and then suddenly jump to the other state in about three call holding times as shown in Figure 6.35.b.

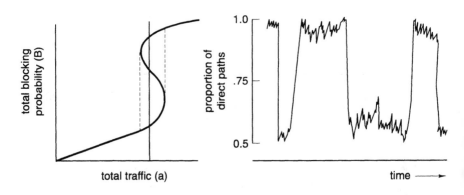

Figure 6.35   Dynamic behaviour of mutual overflow system

Clearly, this is undesirable behaviour in a real network, and precautions, such as trunk reservation, would be taken in practice to prevent such instability.

6.12 References

1   GIRARD, A.: 'Routing and dimensioning of circuit-switched networks', Addison-Wesley, Reading MA (1990)

2   LITTLECHILD, S. C.: 'Elements of telecommunications economics', Peter Peregrinus Ltd., on behalf of the IEE (1979)

3   MORGAN, T. J.: 'Telecommunications economics', (2nd edition), Technicopy Ltd, Stonehouse, Glos. (1976)

4   CATTERMOLE, K. C and O'REILLY J. J.: 'Problems of randomness in communication engineering', Pentech Press (1944)

5   LEON-GARCIA, A.: 'Probability and random processes for electrical engineering', Addison-Wesley (1989)

6   ALLEN, A. O.: 'Probability, statistics, and queuing theory with computer science applications', Academic Press (1978)

7   TAKAGI, K.: 'Optimum channel graph of link systems' Electronics and Communication in Japan, 54A, No 8, (1971) pp. 1–10

8   BEAR, D. 'Principles of telecommunication traffic engineering', (Rev. 3rd edition), Peter Peregrinus Ltd, on behalf of the IEE (1988)

Chapter 7

Packet networks

7.1 Messages and packets

Message switched systems have been known since the early days of telegraphy. The simplest system was one in which telegrams or messages were received over a link in the form of punched tapes in the Morse or teleprinter code. The tape could then be used in a transmitter which sent the message over the next link towards its destination.

In message switching, each switching node stores the incoming message and then forwards it over the outgoing route. Long messages in such a store-and-forward switching strategy can impose large queuing delay on other users sharing the route. Thus, in contrast to circuit switching, where in the event of heavy traffic, calls are blocked, in message switching calls face long delays. To reduce delay, a message can be split into several short notes, called packets. In this case, before a complete message from one source has been fully routed, other users can interleave parts of their messages in packet forms. Hence, with packet switching the queuing delay is more fairly distributed among the users sharing the switching node.

Packet switching can be connectionless or connection-oriented. In connectionless switching, like message switching, packets are stored and forwarded. This technique is also called datagram. The outgoing route can change from time to time, depending on the routing strategy. Hence packets may arrive out of sequence. Reordering of the packets to be assembled into the original message may cause additional delay. In contrast, in connection-oriented switching, all packets belonging to a message always follow the same route, so avoiding the out of sequence problem. In this case, prior to data transfer, a link between the source and destination is set up. This is similar to the call set up phase in circuit switching. Also to minimise overhead information, only an abbreviated 'address' need be given to obtain the full address and routing instructions, which are stored at each node when the connection is set up.

Since in packet switching (connectionless or connection oriented) individual links are temporarily assigned to a packet, each packet then needs a header to specify its destination address. The header might also include some additional bits for, say: packet sequence number, error correction, packet length, etc. It might appear that these extra overheads will reduce the transmission efficiency of packet switching in comparison to circuit switching, where no additional information (apart from the call set up signalling) is added to the actual data. However, this depends on the characteristics of the data to be transmitted. For example, in communication between remote terminals and a computer, the speed of generating commands can be much lower than the data rate of the link. The data in this case is very bursty. The link is busy only for a very short period of time, and is silent for a long period. If a permanent connection, as in circuit switching, is used, there is a significant waste of channel capacity. With packet switching, during the silence period of one terminal, others can send their data, and hence the channel is better utilised. In general, all kinds of bursty traffic, data, or even compressed speech and video can benefit from packet switching. However, since interactive speech and video are delay sensitive services, they cannot tolerate long delays, nor any delay variation. In this case small fixed-size packets are usually employed. The asynchronous transfer mode (ATM) is a type of fast packet switched transport, which employs small fixed length packets (called cells). Demand for transmission of real-time services as well as data through ATM networks has opened up the opportunity that a single broadband network can be used for the transfer of all kinds of services. Currently worldwide research on the characteristics of ATM networks is in progress.

ATM is also a connection-oriented system, and with fixed size cells always following the same route, there is a substantial correlation between the cell arrival rates. This complicates the performance analysis of the network, as incoming cells are no longer independent of each other. To some extent, connection-oriented packet switching, even with variable size packets, might also exhibit residual correlations between the incoming packets. Throughout this chapter we assume connectionless packet/message switching systems such that incoming packets can be assumed to be independent. Furthermore, for variable length packets, in particular varying with an exponential distribution, one can assume a random service time. This makes network analysis very simple, since now the network can be modelled with an M/M/1 queuing process.

7.2   Data link as a queue

Chapter 4 dealt in general terms with the analysis of the performance of queues. In this chapter we extend the theory to deal specifically with queues related to message/packet networks.

A simple data link may be considered as a queue, as shown in Figure 7.1.

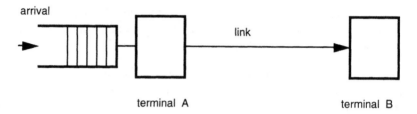

terminal A   terminal B

Figure 7. 1 Simple message or packet link

Packets of mean length **m** bits arrive at a rate λ s^{-1} for transmission over the link, which has a capacity of **C** bit/s.   The service rate μ is then equal to

$$\mu = \frac{C}{m}$$

(7.1)

and the average service time of the link is $1/\mu$. The average delay time, τ, that the data experience in going from terminal **A** to terminal **B**, is the sum of the average waiting time in the queue, **w**, and the average service time;

$$\tau = w + \frac{1}{\mu}$$

(7.2)

In chapter 4 it was shown that the average waiting time for a M/M/1 queue is

$$w = \frac{\frac{\lambda}{\mu}}{\mu - \lambda}$$

(7.3)

Hence, the mean message delay on the link is

$$\tau = w + \frac{1}{\mu} = \frac{1}{\mu - \lambda}$$

(7.4)

and the total link delay per unit time is

$$\tau_{tot} = \lambda\tau = \frac{1}{\mu - \lambda}$$

$$(7.5)$$

7.3   Poisson and near-Poisson queues

Although Poisson queues have been discussed in Chapter 4, it is worthwhile to summarise the main properties in connection with packet switched networks. First, we assume that the system is in statistical equilibrium. As for switched networks, this means that the traffic is assumed to be constant over an interval of time (e.g. the busy hour) that is much greater than the interval between events such as packet arrivals. Next it is assumed that the arrival process is Poisson, that is arrivals represent independent events and, as has been shown, the interarrival times have a negative exponential distribution.

The assumption of a pure Poisson process assists analysis since the sum of a number of independent Poisson processes is also Poisson [1]. In a large network, the packets arrive at a node from many sources and may therefore be considered to be independent. In fact, the sum of the packets arriving at a node as a result of many stationary point processes (not necessarily Poisson) tends asymptotically to a Poisson process.

For a message system the service times may usually be assumed to be independent and to have a negative exponential distribution, since the short messages usually tend to predominate and messages have decreasing probabilities of occurrence as they get longer. This does not necessarily apply to packet systems in which messages are divided into packets. Long messages will be divided into maximum length packets to give a 'blip' in the distribution. However, in most practical packet systems, the assumption of negative exponential service times is a sufficiently accurate approximation.

For a Poisson arrival process and negative exponential service times, it follows that the departure process is also Poisson. However, for a stationary, non-Poisson arrival process and negative exponential service times, the departure process may be shown to be a stationary process that is nearer to Poisson than that of the arrival process.

7.4   Pooled point processes

Consider packets arriving at a node from a number of sources or tributaries. The arrival processes are pooled in the node as shown in Figure 7.2 .

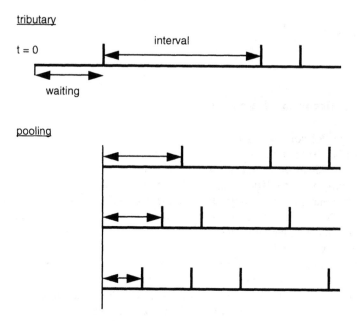

Figure 7. 2   Timing for pooled process

Let us assume that for each source the interval between successive packet arrivals
has a probability distribution denoted by f(t) and the corresponding cumulative
distribution is F(t). Also the waiting time for each arrival has a probability
distribution g(t) and a cumulative distribution G(t). Consider an interval of length x
for one of these sources, as shown in Figure 7.3.

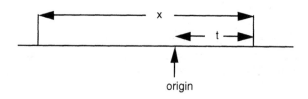

Figure 7.3 Pooling of a new packet within an interval

The origin of a new packet generated within this interval, which is equally likely
to occur at any time, has a conditional waiting probability density function

$$g(t|x) = \frac{1}{x} \quad t < x$$
$$= 0 \quad t \geq x$$

$$(7.6)$$

since within the interval x, each possible origin is equally likely. Hence $g(t|x)=1/x$, and the probability of waiting more than x in an interval of length x is clearly zero.

The probability of choosing an interval of length x is

$$p(x) = xf(x)$$

since the probability is proportional to the probability of an interval being of length x, i.e. $f(x)$, and also to the length of the interval x. Clearly the bigger an interval is, the more likely an arbitrary origin has of falling within it.

Then the unconditional waiting density function is

$$g(t) = \int_0^\infty g(t|x)p(x)dx$$

$$= \int_t^\infty \frac{1}{x}xf(x)dx$$

$$= \int_t^\infty f(x)dx$$

Thus

$$g(t) = 1 - F(t)$$

or

$$\frac{d}{dt}G(t) = 1 - F(t)$$

$$(7.7)$$

For n independent sources with a cumulative waiting function of $G_n(t)$, the waiting time for a packet to be pooled is the waiting time for any of the sources to generate a packet [2]. That is, for the waiting time of the pooled process to be larger than a time period t, the waiting time of all tributaries must be larger than this period, thus

$$1 - G_n(t) = [1 - G(t)]^n$$

$$(7.8)$$

For these sources the relationship between the waiting time density function, $g_n(t)$, and the cumulative arrival function, $F_n(t)$, is similar to that of the single source of Equation 7.7; i.e.

$$1 - F_n(t) = \frac{d}{dt} G_n(t)$$

(7.9)

Substituting for $G_n(t)$ from Equation 7.8 ,

$$1 - F_n(t) = n[1 - G(t)]^{n-1} \frac{d}{dt} G(t)$$

substituting for $\frac{d}{dt} G(t)$ from Equation 7.9

$$1 - F_n(t) = n[1 - G(t)]^{n-1}[1 - F(t)]$$

(7.10)

Figures 7.4 and 7.5 show the complementary cumulative (tail) waiting and arrival times for different numbers of sources in the pooling process. Both figures clearly show that, even for constant length packets, as the number of tributaries increase, both the waiting time and interval tails tend to negative exponential.

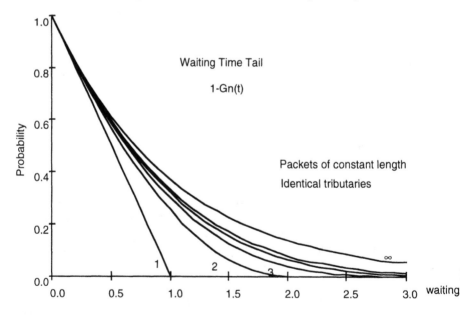

Figure 7.4 Tails of waiting time distributions

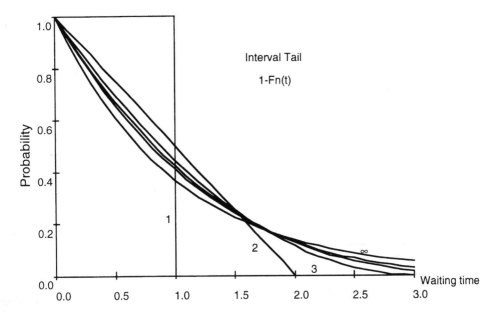

Figure 7.5 Tails of interval distributions

The coefficient of variation (the ratio of standard deviation to mean) of the inter-arrival times is strongly influenced by the type of model used as shown in Figure 7.6. For uniform arrivals, there is no variation and so $C = 0$. If the arrivals are random, the negative exponential distribution of inter-arrival time gives $C = 1$ irrespective of the number of tributaries. However, if we consider an interarrival process for packetised speech such as described in Chapter 3, the burstiness of the arrival process results in a high coefficient of variation for a small number of tributaries but falling to a value close to unity for ten or more tributaries.

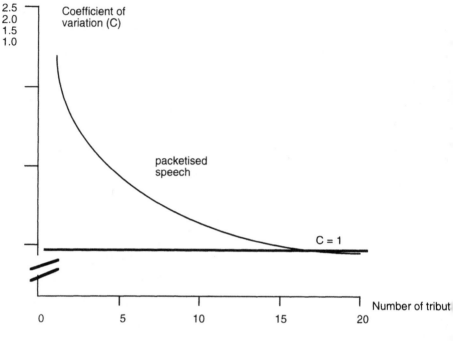

Figure 7.6 Coefficient of variation of inter-arrival time

7.5   Kleinrock's network delay bound and optimisation

Consider a generalised network consisting of links having respective capacities C_i and arrival rates λ_i. Let the messages sent over the link have a negative exponential length distribution with a mean length of m bits. Hence the number of messages sent in unit time is $\mu = C_i/m$. For the exponential distribution the mean holding time is equal to m/C_i. Hence the mean message delay on the ith link is

$$\tau_i = \frac{1}{\dfrac{C_i}{m} - \lambda_i}$$

(7.11)

and the total network delay per unit time is

$$S = \sum_i \lambda_i \tau_i = \sum_i \frac{\lambda_i}{\dfrac{C_i}{m} - \lambda_i}$$

(7.12)

The aim is to reduce the total network delay per unit time. This can, of course, be achieved by increasing the individual circuit capacities, C_i. However, it is more demanding to minimise the total delay with the constraint that the total channel capacity and the arrival rate are fixed. Let us assume the total available channel capacity is C, with

$$C = \sum_i C_i$$

and the total network arrival rate λ, with

$$\lambda = \sum_i \lambda_i$$

It is then possible to assign the link capacities so that the total network delay per unit time, S, is minimum [3]. This is done by differentiating the delay with respect to the individual channel rates, C_i, i.e.

$$\frac{\delta S}{\delta C_i} = \frac{-\dfrac{\lambda_i}{m}}{(\dfrac{C_i}{m} - \lambda_i)^2}$$

(7.13)

Now consider the balance between any two links i,j. For a total fixed channel capacity, an infinitesimal increase in capacity dh in link i is accompanied by a similar decrease in capacity in link j so that

$$dh = dC_i = -dC_j \qquad \text{for } i \neq j$$

(7.14)

Hence the partial individual derivative is

$$\frac{\delta S}{\delta h} = \frac{-\dfrac{\lambda_i}{m}}{(\dfrac{C_i}{m} - \lambda_i)^2} + \frac{-\dfrac{\lambda_j}{m}}{(\dfrac{C_j}{m} - \lambda_j)^2}$$

(7.15)

The turning point is where each individual term is constant irrespective of the channel; i.e.

$$\frac{\lambda_i}{(\frac{C_i}{m} - \lambda_i)^2} = K$$

(7.16)

where K = constant

then

$$\frac{C_i}{m} - \lambda_i = (\frac{\lambda_i}{K})^{\frac{1}{2}} = \frac{\sqrt{\lambda_i}}{\sqrt{K}}$$

(7.17)

Summing for the channels,

$$\sum(\frac{C_i}{m} - \lambda_i) = \sum\sqrt{\frac{\lambda_i}{K}}$$

(7.18)

$$\sum\frac{C_i}{m} - \sum\lambda_i = \frac{\sum\sqrt{\lambda_i}}{\sqrt{K}}$$

Substituting for the total channel capacity, C, and the offered traffic, λ

$$\frac{C}{m} - \lambda = \frac{\sum\sqrt{\lambda_i}}{\sqrt{K}}$$

(7.19)

Hence K can be found in terms of total and individual channel capacity and offered traffic;

$$K = \left(\frac{\sum\sqrt{\lambda_i}}{\frac{C}{m} - \lambda}\right)^2$$

(7.20)

Combining Equations 7.18 and 7.20, the total network delay per unit time is

$$S = \frac{m\{\sum \sqrt{\lambda_i}\}^2}{C - m\lambda}$$

(7.21)

Since

$$\{\sum_{i=1}^{N} \sqrt{\lambda_i}\}^{\frac{1}{2}} > \{\sum_{j=1}^{M} \sqrt{\lambda_j}\}^{\frac{1}{2}}$$

(7.22)

when

$$\sum_{i=1}^{N} \lambda_i = \sum_{j=1}^{M} \lambda_j = \lambda \quad \text{and} \quad N > M$$

(7.23)

then the total network delay per unit time, S, can be reduced if the number of links for a given level of offered traffic is reduced. This implies merging some short links into longer ones. However, longer routes also tend to increase the average delay. This is because if the total mean message entry is ($\gamma \leq \lambda$), the mean path length, L, is given by

$$L = \frac{\lambda}{\gamma}$$

Then the mean message delay is

$$\tau = \frac{S}{\gamma} \geq \frac{mL}{C - m\gamma L} \{\sum_i \sqrt{\frac{\lambda_i}{\lambda}}\}^2$$

(7.24)

which indicates the mean delay increases with the length. This appears to be in contradiction with the previous remarks, and the best compromise might be a network with not many alternate routes, but each as short as possible.

Example:
Consider two links, A–B and B–C. in a network The mean arrival rate on link A–B, $\lambda_1 = 0.5$ and on link B–C $\lambda_2 = 0.9$ packets per second. The message entries are

$A–B$ γ_1 = 0.2; $B–C$ γ_2 = 0.6 and $A–C$ γ_3 = 0.3 s.

Find the mean message delay if
(a) the links have equal capacities (each = 1 packet s^{-1})
(b) the capacities of the links (total = 2 packet s^{-1}) are
adjusted to give minimum delay.

(a)

total capacity	C = 2	packet s^{-1}
mean	m = 1	packets
total arrival rate	λ = 1.4	packet s^{-1}

Then the network delay per unit time is

$$S = \frac{0.5}{1-0.5} + \frac{0.9}{1-0.9} = 10$$

and the total mean message entry is

$$\gamma = 0.2 + 0.6 + 0.3 = 1.1$$

Mean message delay (s)

$$\tau = \frac{S}{\gamma} = \frac{10}{1.1} = 9.09$$

(b) from Equation 7.17,

$$\frac{C_1}{m} = 0.5 + 0.6\frac{\sqrt{0.5}}{\sqrt{0.5} + \sqrt{0.9}}$$

$$= 0.756$$

and

$$\frac{C_2}{m} = 0.9 + 0.6\frac{\sqrt{0.9}}{\sqrt{0.5} + \sqrt{0.9}} = 1.244$$

Then

$$S = \frac{0.5}{0.756 - 0.5} + \frac{0.9}{1.244 - 0.9} = 4.569$$

and mean message delay (s)

$$\tau = \frac{4.569}{1.1} = 4.15$$

Note that if we adopted a more realistic, quantised allocation of link capacities such that the A–B link had a capacity of 2/3 of the mean and the B–C link 4/3 of the mean, the mean delay rises to only 4.61. Thus the optimum allocation of link capacities is fairly flat but the mean delay rises steeply when the allocation is far from the optimum.

7.6 Network delay and routing

It is often found that the connection between two points in a network is served by two or more multiplexed links. The criteria used for optimally dividing the traffic between the links could be either minimisation of the total delay or maximisation of the network utilisation. While network operators favour maximum utilisation, users would prefer minimisation of the total delay. In general, three strategies are possible for dividing the traffic between the links, each one giving different delay statistics.

(i)   Fixed or static routing: the approach in this case is to calculate the traffic on each link and then to calculate the link delays from M/M/1 queuing theory (Chapter 4). The end-to-end delay is then the sum of the link delays. The optimal routes are the ones that minimise the delay, with the constraint that the total channel capacity is fixed. This is the Kleinrock network delay bound, the optimisation method for which was described in the previous section. It was concluded that the total delay is minimised by reducing the number of switching nodes. However, this might increase the length of certain routes, which also increases delay. Therefore, the optimal routes are the ones that lead to a smaller number of switching nodes with shortest length.

(ii)   Random split: the traffic is split between the routes according to fixed parameters or on the basis of long-term adaptation. The delays are calculated as in (i) and an attempt is made to find a near optimal division of traffic.

(iii)   Short term adaptation: traffic is split between the routes in accordance with the current state of the network. The simplest example of such adaptation is for each message to join the shortest queue. A product-form solution is not possible since the arrivals at a queue are no longer independent.

7.7   Finite buffer capacity

7. 7.1   *Uniform loss network (hypothetical)*

Consider the network shown in Figure 7.7.

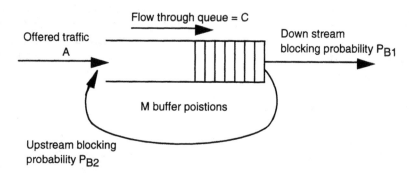

Figure 7.7   A hypothetical queue with finite buffer

The queue contains a limited number of buffer positions (M), the offered traffic is A and the flow through the queue is C. The probability of a message or packet being blocked downstream is p_{B1}. Then the offered and carried traffic are related by

$$A = C(1 - p_{B1})$$

(7.25)

If the message is blocked it is recycled to join the tail of the queue. Because of the limited number of buffer positions, there is a probability p_{B2} that the message will be blocked upstream. This is the loss probability of a M/M/1 queue, shown in Section 4.8 of Chapter 4. Considering that in the figure, there are M buffer positions, then the number of states of the system N = M+1. Thus the upstream blocking probability p_{B2} with N states is;

$$p_{B2} = \frac{C^N(1-C)}{1 - C^{N+1}}$$

(7.26)

If we make the assumption that the upstream and downstream blocking probabilities are similar and equal to p_B, Equations 7.25 and 7.26 may be solved to give the mean throughput, T

$$T = A(1 - p_B)$$

(7.27)

This is plotted in Figure 7.8. with the number of system states as a parameter. It will be seen that because the system has to hold messages in finite buffers the effect of downstream blocking is to impose a limit on the throughput. Any further increase in offered traffic above this limit causes a downturn in the throughput.

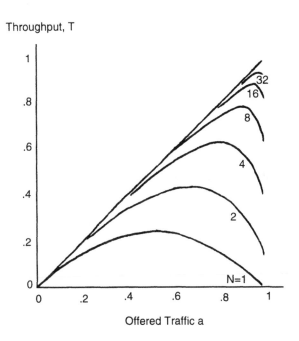

Figure 7.8 Throughput with finite buffer capacity

Although the assumption of uniform blocking is unlikely to represent a condition that occurs in practice, it is useful in illustrating the effects of downturn.

For maximum throughput, let C tend to unity (i.e. $C = 1 - \varepsilon$, $\varepsilon \approx 0$). Then from Equation 7.29, the upstream blocking probability, p_B is

$$p_B \approx \frac{(1-N\varepsilon)-(1-(N+1)\varepsilon)}{1-(1-(N+1)\varepsilon)} = \frac{1}{N+1}$$

(7.28)

Now from Equations 7.25 and 7.28, the offered traffic A is

$$A = C(1-p_B) = \frac{CN}{N+1}$$

(7.29)

At maximum throughput $C \approx 1$ the offered traffic A becomes

$$A = \frac{N}{N+1}$$

(7.30)

and the maximum throughput T is

$$T = A(1-p_B) = (\frac{N}{N+1})^2$$

(7.31)

The expected number of messages in the system is

$$E(n) = \sum_{n=0}^{N} np_n$$

where

$$p_n = p_B = \frac{1}{N+1}$$

Hence

$$E(n) = \sum_{n=0}^{N} n\frac{1}{N+1} = \tfrac{1}{2}N(N+1)\frac{1}{N+1} = \tfrac{1}{2}N$$

(7.32)

and the expected delay is

$$E(w) = \tfrac{1}{2}N \quad \text{as } C \to 1$$

(7.33)

7.8 Packet length

Recall that every packet has to prepend a header onto its payload for proper routing and addressing. Therefore short packets can lead to inefficient network utilisation, reducing the data throughput. On the other hand, if packets are made too long, they are more likely to be received in error, which also reduces the throughput. Hence, there should exist an optimum packet length, where the throughput is maximised.

The optimum packet length can depend on several factors. These may include: the number of bits in the header; the protocol used in packet transmission; the channel error characteristics, etc. For simple analysis, let us define the data throughput as the proportion of the safely-received packets (we will see later that this is the throughput of select-repeat protocol, resulting in the best throughput). We also assume that the independence of channel error follows a Poisson distribution with a mean bit error ratio (BER) of p. Therefore the throughput, T, for packets of length x bits including a header of h bits is given by:

$$T = (1 - \frac{h}{x})p_0$$

$$(7.34)$$

Where p_0 is the probability of receiving packets with zero error (safely-received ones). Therefore

$$p_0 = e^{-px}$$

and hence the data throughput is

$$T = (1 - \frac{h}{x})e^{-px}$$

$$(7.35)$$

If the BER, p, is small such that $px \ll 1$, then e^{-px} can be approximated by:

$$p_0 = e^{-px} \approx 1 - px$$

and the throughput of correct packets is

$$(1 - \frac{h}{x})e^{-px} \approx (1 - \frac{h}{x})(1 - px)$$

$$(7.36)$$

To find the packet length that optimises the throughput, we can take the throughput derivative with respect to packet length and set it to zero.

$$T = (1 - \frac{h}{x})(1 - px)$$

$$\frac{dT}{dx} = (1 - \frac{h}{x})(-p) + (1 - px)\frac{h}{x^2}$$

$$\frac{dT}{dx} = 0 \Rightarrow p(1 - \frac{h}{x}) = (1 - px)\frac{h}{x^2}$$

$$\Rightarrow x^2 = \frac{h}{p}$$

$$\Rightarrow x = \sqrt{\frac{h}{p}}$$

(7.37)

Figure 7.9. shows the packet throughput versus the packet length, x, for a given header size of $h = 48$ bits and various error rates.

In the figure the dashed lines are the throughput derived using the exact Equation 7.40.

It is seen that for small error rates such as experienced on optical fibre links, the packet length is not very critical. For links with high error rates, the optimum packet size is in the order of 1 kbits. In practice lengths of this order have been used. For example ARPANET uses $x = 1008$ bits and British PSS uses $x = 1096$ bits [4, 5]. Note that the optimum length also depends on the header size as well as the protocol used for packet transmission. These protocols are discussed in the following section.

7.9   Data link repeat protocol

To protect packets against channel errors, any attempt at forward error correction of relatively long packets of the order of 1 kbits, can significantly reduce the data throughput. Instead, packets are simply detected for possible error and they are automatically requested for retransmission. This is called automatic repeat request (ARQ) protocol. A number of ARQ procedures are possible [6]. In the following, the throughput performance analysis of those ARQs used in practice are made. In all cases it is assumed that the sender always has packets ready for transmission. This of course gives an upper bound on the performance.

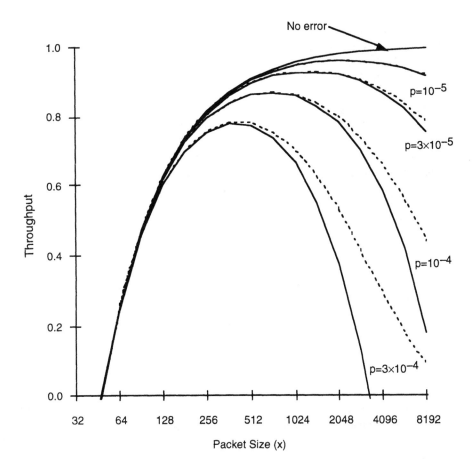

Figure 7.9 Packet throughput versus packet length

7.9.1 Stop-and-wait protocol

In this method, the sender expects an acknowledgement from the receiver for every packet sent. However, if the receiver fails to acknowledge, after a time-out, the sender repeats the unacknowledged packet. The throughput of a system using such a protocol, is heavily influenced by the the round trip delay between the sender and receiver, in addition to channel errors. We also assume that for maximum possible throughput the time-out can be made equal to this delay.

In a channel with the capacity of C bits, the time required for the sender to generate a packet of size x bits is $t_l = x/C$. After a one-way propagation delay of t_p, the receiver has to process the packet for admission, requiring a processing time of t_{prc}. Finally, the receiver has to reply by an acknowledgment packet, which will reach the sender after another propagation delay, t_p. For increased transmission efficiency, the acknowledgement packets can be as small as a packet header with h bits, hence requiring $t_a= h/C$ s. Therefore the shortest time interval between successive packet transmissions, t_T, is

$$t_T = t_l + 2t_p + t_{prc} + t_a$$

(7.38)

However, due to channel errors, erroneous packets have to be retransmitted. This makes the average interval between useful packet transmissions longer than t_T. If we assume a packet is in error with a probability of q, where

$$q = 1 - e^{-px}$$

then the average time for a correct transmission, t_v, is

$$t_v = t_T + (1-q)\sum_{i=1}^{\infty} iq^i t_T$$

$$= \frac{t_T}{1-q}$$

(7.39)

Let us assume the ratio of the shortest time interval, t_T to the time taken to generate a packet, t_l, to be a, i.e $a=t_T/t_l$, then the data throughput is the useful proportion of the average time for a correct transmission, t_v, i.e.

$$T = (1 - \frac{h}{x}) \frac{t_l}{t_v}$$

and substituting for t_v

$$T = (1 - \frac{h}{x}) \frac{1-q}{a}$$

(7.40)

7.9.2 Go-back-N

Due to its simplicity the stop-and-wait protocol is ideal for half-duplex transmission on short routes. However, as the path length increases, so does the propagation delay, t_p, and as a result a increases, reducing the data throughput. For example on a satellite route with $t_p= 240$ ms, for packets of length 1200 bits with a header of 48 bits, in a channel with a capacity of 48 kbits/s, (ignoring the processing time, t_{prc},)

$$t_l = \frac{1200}{48000} = 25\text{ms}$$

and

$$t_T = (2 \times 240) + 25 + \frac{48}{48000} = 506\text{ms}$$

then

$$a = \frac{t_T}{t_l} = 20$$

Thus the ideal throughput of $1-h/x$ is reduced by a factor of 20.

One solution to this inefficiency is the go-back-N strategy. In this method the sender does not wait for the receiver's acknowledgement, and sends packets as they are ready for transmission without waiting for acknowledgement. In case the packet is not acknowledged, the sender has to retransmit all packets following the unacknowledged packet as well as the unacknowledged packet itself. For this purpose, the sender employs a buffer and stores the sequentially numbered sent packets in its buffer for possible retransmission. Figure 7.10 shows an example of this protocol, where after sending packet number 6, the sender either receives a negative acknowledgement, or a time-out warning, indicating that the transmission of packet number 3 was not successful. In this case the sender has to go back, and retransmit packet number 3 and all the following ones, even though they might have been received safely.

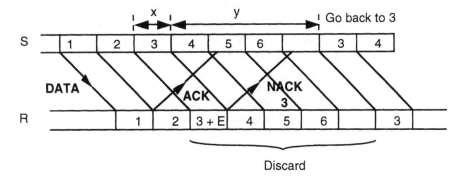

Figure 7.10   Go-back-N data flow

The performance analysis of this protocol can be made in a similar way to the stop-and-wait protocol, with the difference that the average time interval between the successive sent packets is now t_l but that of errors is still t_T, hence the average time for a correct transmission is

$$t_V = t_l + (1-q)\sum_{i=1}^{\infty} iq^i t_T$$

again using $t_T = a\, t_l$;

$$t_V = t_l \times \left[\frac{1+(a-1)q}{1-q}\right]$$

(7.41)

Now the data throughput of this protocol, T, is

$$T = \left(1-\frac{h}{x}\right)\frac{t_l}{t_V}$$
$$= \left(1-\frac{h}{x}\right)\left(\frac{1-q}{1+(a-1)q}\right)$$

(7.42)

Note that in this protocol, in contrast to the stop-and-wait protocol, for low bit error rates and small packet sizes the effect of a can be significantly weakened. For example for packets of length 1 kbits and and a BER of $p = 10^{-6}$, the packet error rate q is almost equal to 0.001, significantly nullifying the (a–1) factor of the throughput equation. However, if the error rate is large, or the packet length is too long, then the packet loss rate, q can be significant. For example, for the same packet length, if the BER was 10^{-4}, then the packet loss rate is only 0.1, not weakening the (a–1) factor significantly. The throughput of this protocol for various packet sizes, propagation delays and channel error rates is shown in Figure 7.11. It is seen how increasing the error rate, reduces the throughput. Also, similar to the ideal case, the packet length is optimum around 1 kbits. For small error rates, the throughput becomes less sensitive to packet length.

The round trip delay 0 ms is the upper bound delay, 20 ms is a typical terrestrial link delay and 500 ms is that of a satellite route at 64 k bit/s.

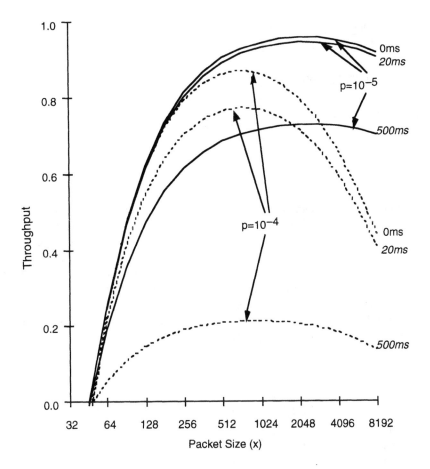

Figure 7.11 Throughput of go-back-N for various round-trip delays

7.9.3 Select-repeat protocol

If in the previous protocol only the erroneous packets are transmitted, then the throughput can be increased further. However, this leads to out of order reception of the packets, and hence the receiver also needs a buffer, at least as large as that of the sender to reorder the packets. The throughput of this protocol can also be analysed in a manner similar to the stop-and-wait and the go-back-N protocols, with the exception that both the average time between the successive packets and the retransmitted ones are t_l. Thus the average time for correct transmission, t_v, is

$$t_v = t_l + (1-q)\sum_{i=1}^{\infty} iq^i t_l$$

$$= \frac{t_l}{1-q}$$

(7.43)

and the data throughput, T, is

$$T = (1-\frac{h}{x})\frac{t_l}{t_v}$$

$$= (1-\frac{h}{x})(1-q)$$

(7.44)

Note that this throughput is in fact that which was used in deriving the optimum packet size. Here the factor **a** has no effect on the throughput, and hence the throughput of this protocol can be better then the go-back-N one. Figure 7.12 compares the performance of these two protocols for a channel error ratio of $p_e=10^{-5}$ bit/s. The channel capacity is assumed to be 48 kbit/s, the packet header h=48 bits and the round-trip delay $t_p=350$ ms. It is seen the relative advantage of select-repeat over go-back-N is marginal. In practice this marginal advantage does not justify the extra complexity required at the receiver for this protocol, and hence practically go-back-N is usually preferred.

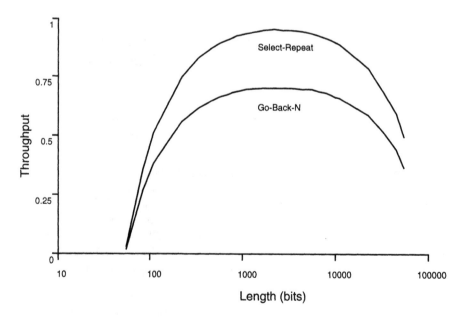

Figure 7.12   Relative efficiency of selective-repeat over go-back-N

7.10   Flow control

Flow control mechanisms are normally used in packet networks to regulate the flow of traffic between the the sender and the receiver. This is needed either to match the speeds of the sender and receiver, or to control congestion and preserve stability in the network. For example in X.25, which is an interface protocol between the data-terminal equipment (DTE) and the data circuit-terminating equipment (DCE) flow control makes sure that the DCE is not exhausted with the flow of data from its DTE [7, 8]. This is not because DCE might be slower than DTE, but its access to a congested network necessitates this.

To control the flow of data, the receiver employs a window, whose size represents the maximum number of packets outstanding in any direction (e.g. in X.25 from DTE to DCE or vice versa). The specific value of the window size is negotiable, and it affects the network performance.

An example of flow control is the sliding window protocol. In this protocol, if the difference between the sequence number of the last packet sent, S, and the sequence number of the last acknowledged packet, R, is less than the window size, N, then the sender is allowed to send new packets, otherwise it stops sending. Figure 7.13 shows the flow of traffic for a sliding window of size $N=4$ packets and maximum sequence number 8. As long as the number of packets in transit $S-R$ is less than or equal the window size, $S-R \leq N$, flow of traffic continues, otherwise it is halted.

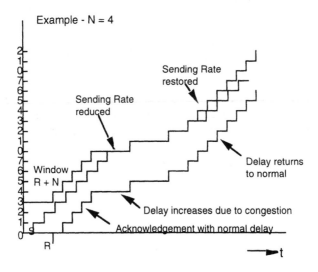

Figure 7.13 An example of sliding window traffic flow control

Note that if the congestion is on the return circuit rather than the go circuit, the packets being sent will also be delayed. In some networks, acknowledgement packets may be given higher priority.

7.10.1   Simple model for flow control analysis

The performance of the sliding window flow control, described in the previous section, can be evaluated by examining the time-delay and throughput of a network under such a flow control mechanism. Let us assume that the circuit connecting a source to its destinations traverses M packet switching nodes. Each node is assumed to be a simple store-and-forward switch, hence the the circuit introduces a delay equal to M packets. It is also assumed that packet lengths at each node are independent of each other and follow a Poisson distribution. The holding time follows an exponential distribution with mean holding time $1/\mu$. So that the service rate of each node is μ. This is in fact an M/M/1 queue. We also assume received packets at the destination are immediately acknowledged. Hence a simple model for such a link is a cascade of M queues, as shown in Figure 7.14

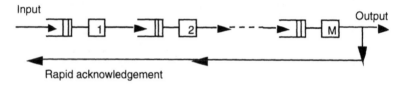

Figure 7.14 A simple queuing model of a link with rapid acknowledgement

Light traffic
For light traffic of n packets in transit and a window size of N, $(n \le N)$, the throughput, $\gamma(n)$, with n packets uniformly distributed among the M queues is the proportion of the service rate, μ, for which a queue is not empty;

$$\gamma(n) = \mu \times \text{prob(a queue is not empty)}$$

For a queue of size M and n packets in transit, the probability is;

$$\text{Prob (a queue is not empty)} = \frac{E(n)}{E(n)+M-1}$$

where $E(n)$ is mean number of packets in transit, and -1 implies that one packet has already been served. Thus the throughput becomes

$$\gamma(n) = \mu \times \frac{E(n)}{E(n) + M - 1}$$

$$(7.45)$$

The mean delay can be calculated by using Little's formula, that states: the mean delay, $E(T)$, through a virtual channel is the ratio of the average number of packets, $E(n)$, through the channel to the channel throughput, $\gamma(n)$, i.e.

$$E(T) = \frac{E(n)}{\gamma(n)}$$

$$(7.46)$$

Substituting for the throughput, $\gamma(n)$, from Equation 7.45 gives

$$E(T) = \frac{E(n) + M - 1}{\mu}$$

or the normalised mean delay becomes

$$\mu E(T) = E(n) + M - 1$$

$$(7.47)$$

The normalised throughput γ/μ and normalised mean delay $\mu E(T)$ can be combined to form what is called the throughput-mean-delay trade off characteristics

$$\mu E(T) = \frac{M - 1}{1 - \frac{\gamma(n)}{\mu}}$$

$$(7.48)$$

Since the normalised throughput γ/μ is usually defined as traffic load a, then the throughput-delay trade off equation is given by

$$\mu E(T) = \frac{M - 1}{1 - a}$$

$$(7.49)$$

Heavy traffic
The throughput and the mean-delay characteristics for the heavy traffic can be derived by assuming that the input traffic λ is much larger than the service rate μ. In such a condition it is very likely that the window is always full. As soon as an

acknowledgment packet is sent, the window slides by one step, and soon becomes full again. Hence the average number of packets in transit is N, i.e. $E(n) = N$. Thus the throughput and mean delay under heavy traffic become:

$$\gamma = \frac{N\mu}{N+M-1}$$

$$(7.50)$$

The normalised throughput a is

$$a = \frac{\gamma}{\mu} = \frac{N}{N+M-1}$$

$$(7.51)$$

and the normalised mean-delay is

$$\mu E(T) = N+M-1 = \frac{M-1}{1-a}$$

$$(7.52)$$

Note that the sliding window with rapid acknowledgement is an ideal case, which result in a maximum throughput and minimum mean delay. Thus the given normalised throughput and mean-delay are the upper bound for the performance of a practical window flow control mechanism. Hence in practice, the normalised throughput and mean delay are:

$$a \leq \frac{N}{N+M-1}$$

$$(7.53)$$

and

$$\mu E(T) \leq \frac{M-1}{1-a}$$

$$(7.54)$$

Figure 7.15 shows the normalised mean delay bound versus traffic load a for two buffer sizes of $M=3$ and $M=5$.

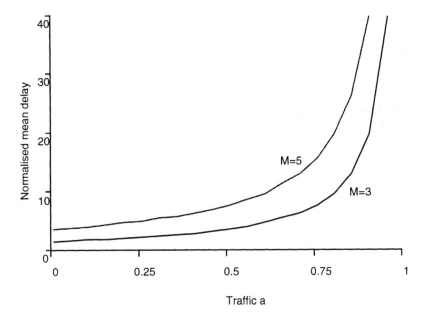

Figure 7.15 Time delay-throughput tradeoff bound

Optimum window size
Both the throughput and mean-delay vary with the window size, N. While the
mean delay increases linearly with N, the increase in throughput is non-linear. The
throughput grows very quickly with N, but saturates to the maximum service rate,
μ. It is desired to increase the throughput, but to minimise the mean-delay time.
This can be achieved by joint optimisation of the product of the normalised
throughput and the inverse mean-delay, i.e.

$$\frac{\delta}{\delta N}\left(\frac{\gamma}{\mu} \times \frac{1}{\mu E(T)}\right) = 0$$

$$(7.55)$$

Substituting for the normalised throughput and mean-delay, then,

$$\frac{\delta}{\delta N}\left(\frac{N}{(N+M-1)^2}\right) = 0$$

which results in

$$(N+M-1)^2 = 2N(N+M-1)$$

or

$$N = M - 1$$

(7.56)

Note again that this is an optimum size for a rapid acknowledgement. In practice the 'ack' packets also experience delay (see later), which is equivalent to the packets crossing through more nodes. Hence N is larger than M and typically it lies in the range of M to 3M.

7.10.2   *Parallel virtual circuits with flow control*

In the previous section we looked at a simple network consisting of only one virtual circuit, VC. In practice a network of M switching nodes may be shared by several VCs. Each virtual circuit, VC_i, can have its own number of packets in transition, n_i, and its own window size, N_i. VCs tend to impose delay on each other hence affecting throughput and mean-delay. Figure 7.16 shows a queuing model for the parallel virtual circuits passing through M switching nodes. Again it is assumed that the service rates of all nodes are identical and equal to the inverse mean holding time, μ. We also assume that the acknowledgment is rapid, hence the derived network characteristics represent the upper bound to the performance.

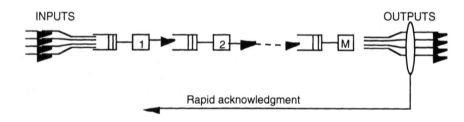

Figure 7.16   A queue model for parallel virtual circuits

As in the case of the analysis of a single VC, each individual VC encounters a delay of M queues plus the sum of packets in transit. That is, the normalised mean-delay of the ith VC, VC_i, is bounded to

$$\mu E(T_i) \le M + \sum E(n_i) - 1$$

(7.57)

which is the same for all VCs. However, the normalised throughputs γ of the circuits are different and are proportional to their mean numbers of packets in transit. The throughput of the ith circuit is then:

$$a = \frac{\gamma_i}{\mu} \le \frac{E(n_i)}{M-1+\sum E(n_i)}$$

(7.58)

and the total traffic, a, is

$$a = \sum a_i \le \frac{\sum E(n_i)}{M-1+\sum E(n_i)}$$

(7.59)

For heavy traffic, where each window is always full, i.e. $E(n_i) = N_i$, then the individual traffic and the total traffic are respectively;

$$a_i \le \frac{N_i}{M-1+\sum N_i}$$

and

$$a = \sum a_i \le \frac{\sum N_i}{M-1+\sum N_i}$$

(7.60)

Their ratio is then

$$\frac{a_i}{a} \le \frac{N_i}{\sum N_i}$$

(7.61)

This implies that the portion of the total traffic for each VC, is the ratio of its saturated packets in transit (window size) to the total traffic. However, all VCs experience the same delay, that can be written in terms of total traffic, a, as

$$\mu E(T) = \frac{M-1}{1-a}$$

(7.62)

The total window size in the parallel route is

$$N = \sum N_i$$

Hence the total traffic becomes

$$a = \frac{N}{N+M-1}$$

(7.63)

Equations 7.62 and 7.63 for the parallel VCs are similar to those derived for a single VC. Thus a route with parallel VCs acts like a single VC carrying the same total traffic and has a window size equal to sum of the individual window sizes. The mean average delay (equal for all the VCs) is given by Equation 7.62, derived from the total traffic, **a**.

In principle under all loads, both the mean delay and the overall throughput are stable, but for large N, random fluctuations are possible. For example if the bursts of traffic from several VCs do overlap, then both the mean-delay and throughput are impaired. When r bursts overlap, then

$$\sum E(n_i) = rN$$

(7.64)

and hence the normalised mean delay and throughput of each VC become

$$\mu E(T) = M - 1 + rN$$

and

$$a_i = \frac{N}{M-1+rN}$$

(7.65)

The total traffic is

$$a = \frac{rN}{M-1+rN}$$

(7.66)

The effect of overlapped bursts on the network characteristics is tabulated in Table 7.1 for a M=3 node link having a total window size of 2.

Table 7.1 Numerical example with M=3 and N=2

r	a_i	a	$\mu E(T)$
1	0.500	0.500	4
2	0.333	0.667	6
3	0.250	0.750	8
4	0.200	0.800	10
5	0.167	0.833	12

Figure 7.17 shows that with overlapped bursts, no one user can 'freeze out' the other users. This is a simple form of capacity sharing.

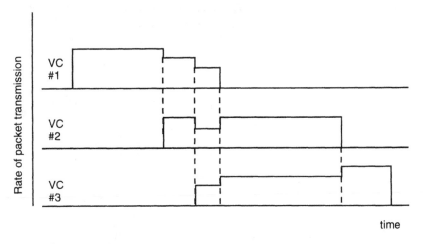

Figure 7.17 Overlapped bursts of traffic on parallel VCs

7.10.3   Flow control: further factors

As mentioned in the simple queuing model of a single virtual channel, the acknowledgment of packets may not be rapid. It is more likely that the 'ack' packets in the return route also have to traverse several switching nodes. Hence a more realistic model for a single virtual channel is a symmetric queue as shown in Figure 7.18.

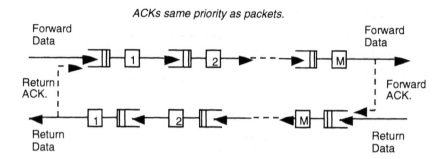

Figure 7.18   Single channel queue model with queued acknowledgement

The length of the chain is effectively doubled to **2M**, and hence an optimum window size is $N \approx 2M$. Furthermore for long routes the propagation delay, τ, can be significant. In this case to keep up traffic flow, the window size has also to be increased by $2\,\mu\tau$.

A further difference between the practical networks and the simple models used here is that all the VCs may not share the whole route. They might only share part of a path, as shown in Figure 7.19, where only node **X** is shared between two VCs. They also can have different path length.

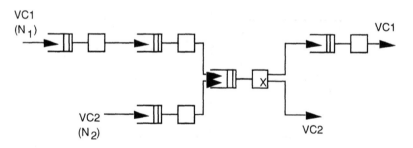

Figure 7.19 Multiple channel queuing model

Since a long distance path (more switching nodes, M=large), necessitates a larger window (e.g. N = 2M), then the window sizes for the VCs can be different. If the shared part (e.g. node **X** in Figure 7.19) is congested, then capacity is shared between the VCs is in proportion to their window sizes, $N_1 : N_2$.

Several methods can be employed to improve network performance. For example, under severe congestion the window sizes of all the individual VCs may be reduced, preferably to a common default value, **N**. Alternately, rather than limiting individual tokens for VCs to have the permission to send packets, the total number of packets in the network can be bounded.

7.11 ATM

In Chapter 6 we looked at circuit switched architectures and in this chapter we have considered packet networks. While circuit switches are suitable for delay sensitive constant bit rate services such as telephony, they are not suitable for bursty traffic such as data. Similarly, packet networks, specially with the optimum packet length of the order of 1k-10k bits (see section 7.9) are not suitable for real-time services. For example transmission of **64** kbit/s speech in such a network requires **16 −160** ms packetisation delay, which is not acceptable for conversational purposes when added to the other delays in the network.

Advances in audio and video technology have now made it possible to code these signals at variable bit rates. The bit streams of these data can be bursty like data, but they are still sensitive to delay. Furthermore, with ever growing computer communication necessities, faster packet switching techniques are required. Moreover, with the increasing demand on multimedia communications, it is highly desirable to accommodate all types of services within a unified transmission media. Since neither of the existing circuit nor packet switching technologies can handle multimedia services, then a new type of transporting network is required.

It is now internationally accepted that the transport media for the future broad-band integrated services digital networks (BISDN) is asynchronous transfer mode (ATM) [9]. In this network, data rates as low as a few hundred bits per second for telemetry and up to tens of megabits for high definition television (HDTV) can be transmitted concurrently. Since in this heterogeneous media, burstiness and delay sensitivity do exist, ATM is based on fast packet switching with small fixed size packets, called cells [10]. Each cell is **53** bytes long with a **5**-byte header and **48** bytes payload, transmitted at rates of **155−625** Mbits/s. An ATM channel is slotted into fixed size **53**-bytes slots, and a source wishing to send data, seizes one of these empty slots and convert them into full cells.

As in the case of circuit and packet switched networks, a study of the performance of ATM networks involves an analysis of the behaviour of ATM switches. Nowadays, a large number of ATM switches are commercially available and although they are classified into various switching fabrics, they all have a common problem known as output contention, which can be used in our performance analysis [11].

In general in an ATM switch any cell in the incoming inlet may wish to be connected to any outgoing outlet, as shown in Figure 7.20.

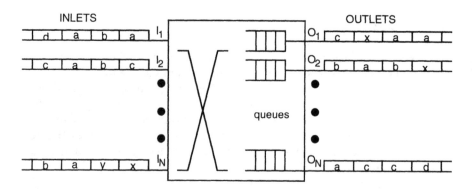

Figure 7.20 An ATM switch

The switching takes place during a time slot equal to the cell size. The switching structure is similar to that of the space-switch used in the circuit switching (Chapter 6). However, the main difference is that in circuit switching the outlet is always available at the appropriate time slot whereas in ATM switches, two or more cells may need to be switched to a particular outlet during the same cell slot. This creates a new problem for ATM switches known as output contention, since at any time several incoming cells may wish to be diverted into one outgoing outlet. ATM switches then may involve queues at the input or the output or both [12]. Output queues occur when the transmission multiplex is temporarily overloaded. Input queues occur when the path through the switch network is temporarily blocked, even though the output link may be free. To avoid excessive input queues, the switch network may be operated at a higher rate than the cell transmission rate or the multiplex. Therefore we have concentrated at the output queue.

As mentioned earlier, in an ATM switch with N inlets and output buffering, switching of every individual cell has to be carried out at N times the cell transmission rate. This allows the switch to divert all the incoming cells to a single outgoing route during a cell slot. Now if we look at one of these outlets, it appears that during one cell transmission/arrival interval, cells from various inlets are multiplexed into this outgoing outlet. Thus the performance analysis of such a switch is similar to that of a single server multiplexing N customers. If we assume that the offered traffic by each customer on any inlet is λ, and each customer is equally likely to chose any outlet, then the offered traffic by each customer to any outlet is λ/N. Thus the probability that k customers out of N are fed to the multiplexer, p_k, is:

$$p_k = \binom{N}{k}(\frac{\lambda}{N})^k (1-\frac{\lambda}{N})^{N-k}$$

$$(7.67)$$

For such a Binomial distribution the average offered traffic by customers in the multiplex is then

$$\text{Load}_{av} = N \times \frac{\lambda}{N} = \lambda$$

which is equal to the incoming and outgoing traffic load.

For sufficiently large N and small λ, Equation 7.67 approximates to a Poisson distribution with parameter λ. This supports our earlier assumptions that cells arriving from a large number of sources have a Poissonian distribution. Also, since the cell size is fixed, the customers are served at constant rate λ. Thus the multiplexer behaviour is similar to that of slotted M/D/1 queue, discussed in Chapter 4. In Section 4.11 it was shown how the state probability of the queue, Q, of such a multiplexer can be calculated. Thus the probability of delay exceeding a certain value of q cells, is simply the cumulative distribution of the state probabilities, Q_k, and is given by

$$p_D(q) = \sum_{k=0}^{q-1} Q_k$$

(7.68)

Figure 7.21 shows the queuing delay for a 150-cell buffer for various multiplex loads. Although the queuing model for load <1 can be effectively modelled with M/D/1/K discrete-state Markov process, analysis for load >1 is difficult [13]. This is reflected in the delay probability distribution function of Figure 7.21, where while for load <1, the distributions are negative exponential, for load >1, they tend towards higher order exponential functions.

Knowing the state probability of the queuing buffer, Q, we can also calculate the mean queue length, Q_{av}, of the queue for a buffer size of B by

$$Q_{av} = \sum_{k=0}^{B-1} k \times Q_k$$

(7.69)

and the mean queuing delay by

$$D_{av} = \sum_{k=0}^{B-1} k \times Q_k \times CTI$$

(7.70)

where CTI is the cell transmission interval.

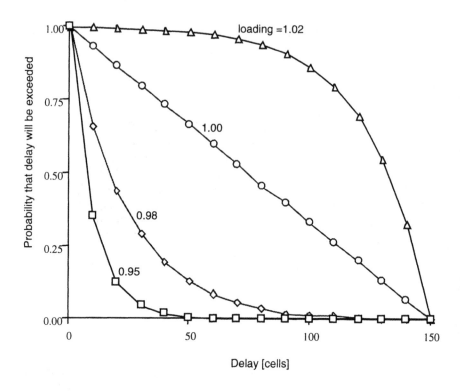

Figure 7.21 Probability of delay exceeding q cells in a buffer of 150 cells

> *Example: a multiplex link can transmit 1000 equal*
> *length packets per second and when no packet is available*
> *an idle packet is transmitted. A buffer store capable of*
> *holding 6 packets is provided at the input to the link.*
> *Packets are generated at random at an average rate of 600*
> *per second. Find the probability that a packet will be*
> *delayed more than 4 milliseconds and the probability of*
> *packet loss due to buffer store overflow.*

The average number of packets generated during a cell transmission interval is 0.6. Then since the packets are generated at random, using the Poisson arrival with a mean value of 0.6 cell per second, the probabilities for the number of packets generated in a cell transmission interval are as given in the following table:

No of packets	Probability
0	0.549
1	0.329
2	0.099
3	0.020
4	0.003

As explained in Chapter 4, by considering each state of the Markov diagram in turn, the statistical stability equations may be set up. For example, if we consider state 2,

$$p_0(0.02) + p_1(0.099) - p_2(0.549+0.099+0.02+0.003) + p_3(0.549) = 0$$

The other equations may be derived in a similar manner and the input matrix for the solution of the equations is:

−0.121	0.549	0	0	0	0	0
0.099	−0.671	0.549	0	0	0	0
0.020	0.099	−0.0671	0.549	0	0	0
0.003	0.020	0.099	−0.671	0.549	0	0
0	0.003	0.020	0.099	−0.671	0.549	0
1	1	1	1	1	1	1

where the final transition equation has been replaced by an equation for the condition that the sum of the state probabilities is equal to unity.

The solution matrix for the equations is then:

state	probability
0	0.731
1	0.162
2	0.067
3	0.026
4	0.010
5	0.004

A packet generated when the model is in states 4 or 5 will experience a delay greater than 4 milliseconds with probability 0.014. The probability of packet loss is approximately 0.0016.

7.12   Modelling traffic sources

In circuit switched systems, it is relatively easy to specify the characteristics of the traffic that is offered to the network. Since a fixed bandwidth channel is held for the duration of a call, it is only necessary to know the statistics of the incidence of calls and of their durations. Packet switching introduces a new element into the performance, particularly if services other than data are involved. Packet networks are variable bit-rate (VBR) systems that offer bandwidth on demand. In other words, the resources of the network are engaged by a source only when packets are to be transmitted. To estimate the performance of the network by analysis and/or simulation, it is therefore necessary to specify both the statistics of the sources generating the information and the rules for assembling the information into packets.

7.12.1   Data sources

Data sources vary greatly from the slow sporadic input from a keyboard to the downloading of large chunks of software such as programmes or computer-aided design information. It is not possible to specify detailed characteristics unless the types of user are known. Furthermore, the method of packetisation varies greatly. Some examples will be discussed in the case study given later in this chapter.

However, in spite of these uncertainties, there is, for practical purposes, a simple way of specifying data source characteristics. Since the average bit-rate of nearly all the individual data sources in a network is much less than the bit-rates of the multiplex links, a high degree of averaging takes place. It is therefore sufficient to model the generation of data packets by a simple Poisson process.

7.12.2   Speech sources

Speech may be digitally encoded into standard PCM or any one of the low bit-rate techniques that have been devised. In most cases, the rate at which bits are generated during speaking intervals ('talkspurts') is constant. During pauses in conversation, usually when listening to the other speaker, there is no need to transmit any information at all. In order to model the speech source, it is only necessary to know the distributions of the durations of the talkspurts and pauses. In some systems, very short talkspurts and pauses (less than about 10-20 ms) are ignored but these have only a minor effect on the results.

The statistics of speech patterns were first measured in connection with time assignment speech interpolation (TASI) systems [14] used for transoceanic cable links. Controlled tests under laboratory conditions [15] provided usable statistical data and later measurements [16] generally confirm the results for English conversation. The on-off patterns of speech suggest that a suitable model might be a simple two-state Markov process shown in Figure 7.22. The transition coefficients have been chosen to fit the measured results in Figure 7.23, with the mean talkspurt duration 1.34 seconds and the mean pause 1.67 s the activity factor is 44%.

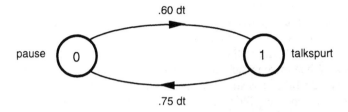

Figure 7. 22   Two-state Markov diagram to model speech patterns

If a number of speech channels are combined in a packet multiplex system, the number of channels active at any one time (proportional to the instantaneous bit-rate) follows a binomial distribution. For a large number of channels this approximates to a normal distribution.

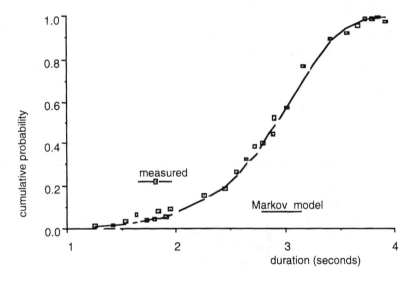

Figure 7.23   Cumulative distributions of talkspurt durations for conversational speech

7.12.3   Video sources

Video coding techniques used to exploit the transmission savings of packet multiplexing are based on the principle that there is no need to transmit the full encoded information for each picture frame. Once the initial scene has been transmitted, only the relatively small updating changes are necessary to reproduce succeeding frames. The resulting bit-rate varies widely. After a scene cut, when the information relating to the complete picture has to be sent, a high bit-rate is necessary. When there is rapid movement in the picture or when the camera pans over a scene the bit-rate is still relatively high. On the other hand, a 'talking head' or a caption requires very little information to enable the next frame to be built up.

Modelling the video source depends on the coding techniques used. This is a rapidly developing area of research and development, although standards (H.261, MPEG2, [17,18] have been produced that enable models of the traffic sources to be devised. However, there remains the problem of what constitutes a 'typical' mix of video sequences. Whereas most telephone conversations are statistically similar in their on-off patterns, a video sequence on a conference TV circuit for example will be very different from that on a broadcast channel that includes action packed sports programmes. Nevertheless, attempts have been made to devise video source models that are sufficiently representative for the performance of a network to be estimated.

The bit-rate profile of a video sequence is shown in Figure 7.24. It will be seen that there is a high degree of autocorrelation in the process and the measured covariance follows an exponentially decreasing function as shown in Figure 7.25 Any model must show a similar autocorrelation effect.

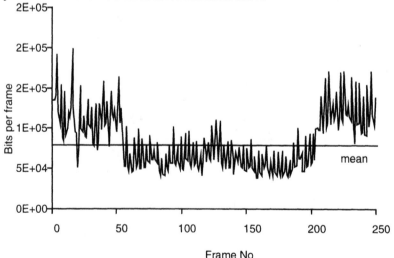

Figure 7.24   *Bit-rate variations for coded video sequence*

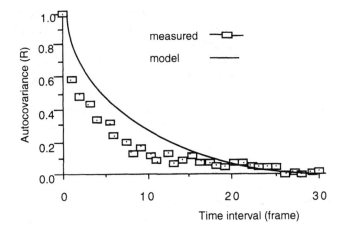

Figure 7.25   Autocovariance for instantaneous bit-rate of video sequence

Autoregressive model
Given an uncorrelated random input ε_n with mean μ_ε and variance σ_ε^2, we can form an auto-regressive process [1] of order 1 by the recursive relation

$$\xi_n = a\xi_{n-1} + b\varepsilon_n$$

(7.71)

If the mean of the input is zero, the mean of the process is also zero and the variance is given by

$$\sigma_\xi^2 = R_\xi(0) = \frac{\sigma_\varepsilon^2}{1-a^2}$$

(7.72)

The autocovariance at time interval τ can then be expressed as a function of the constant a, that is

$$R_\xi(\tau) = R_\xi(0)a^\tau = \frac{\sigma_\varepsilon^2}{1-a^2}a^\tau$$

(7.73)

A video source may then be modelled by choosing the standard deviation σ_ε and the constant a to approximate to the measured bits/frame and autocovariance of the sample video sequences.

For multiplexed signals, it is supposed that an identical autoregressive process is used to model each of the N channels. The bit-rate model for the multiplexed channels is then a similar autoregressive process whose parameters are identical to those of the original process multiplied by N.

Although the autoregressive model is a very simple one for simulation purposes, it is difficult to apply to the analytical evaluation of performance of video packet networks. Furthermore, it is limited to parameters derived from a single 'typical' sequence, and so suffers from being unrepresentative in regions of low probability. Since these are often the most important in the evaluation of a network, the technique should be used with caution.

Discrete-time, discrete state Markov models
A single video source can be represented by a simple M-state Markov model [19, 20] in which the quantised number of bits or packets generated is proportional to the state number (Figure 7.26). Since the statistics of the intra-frame bit-rate are very different from those of the inter-frame, it is convenient to trigger the model at frame intervals (40 ms in Europe and 33 1/3 ms in USA and Japan). Within the frame, the generation of packets is assumed to follow a Poisson process and the average generation rate varies from frame to frame in accordance with the Markov process. The ratios of the transition coefficients into and out of a state are chosen to fit a measured [21] bits/frame distribution while the absolute values fit the autocovariance characteristic.

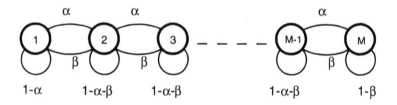

Figure 7.26   An M-state Markov chain modelling a single video source

Since the only transitions permitted are those between adjacent states, the model does not represent the effects produced by scene changes. This may be overcome by postulating additional states, each of which represents the number of bits generated when a scene change occurs and a complete frame has to be built up from scratch.

For a multiplex of N channels, a model of $[N(M-1)+1]$ states may be derived by combining the single source models. Some simplification is possible by appealing to the central-limit theorem and supposing that the transition coefficients out of a particular state have a discrete truncated normal distribution.

Continuous-time, discrete state Markov models
A model developed by Maglaris *et al* [22] also used Markov states to represent the quantised instantaneous bit-rate (Figure 7.27). A is assumed to be a constant depending on the coding process. However, the bit rates are assumed to be sampled at instants which follow a Poisson process and the degree of accuracy whereby the model tracks video signals is determined by the mean sampling rate. The state of the model can change at any time, hence the reference to 'continuous time'. In order to simplify analysis, the model directly represents a multiplex of N channels.

Figure 7.27   A Markov chain representing quantised instantaneous bit-rate

The probability that the model is is state kA is given by the binomial distribution.

$$P_k = \binom{M}{k} \rho^k (1-\rho)^{M-k} \quad \text{where } \rho = \frac{\alpha}{\alpha + \beta}$$

(7.74)

The mean state is given by

$$\mu = MA\rho$$

(7.75)

The variance by

$$Var(0) = MA^2 \rho(1-\rho)$$

(7.76)

and the autocovariance at time interval τ by

$$Var(\tau) = Var(0)e^{-(\alpha+\beta)\tau}$$

(7.77)

Here again the coefficients of the model may be fitted to statistics derived from a video sequence by choosing the ratio $\alpha/(\alpha+\beta)$ to fit the bit-rate distribution and the sum $(\alpha+\beta)$ to fit the exponential constant of the measured autocovariance.

If scene changes have to be taken into consideration a two-dimensional Markov process is required. Sen *et al* [23] extended the earlier continuous-time process to combine a low bit-rate as in Figure 7.27 with a high bit-rate process to represent scenes having rapid movement. This results in a two-dimensional model having NM states. Scene changes are modelled by rapid sampling along the high bit-rate axis. The Markov model appears rather complicated but turns out to be amenable to relatively straightforward analysis.

7.13   Case studies

7.13.1 Case study 1: a token ring

As a first case study we will consider the performance analysis of a token ring local area network architecture (LAN). In a LAN with a token ring topology a group of N stations are attached to a transmission loop (fibre, coax, etc.), as shown in Figure 7.28.

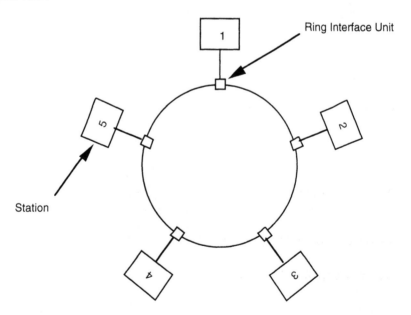

Figure 7.28 A LAN with ring topology

Any station which has permission to send packets seizes an empty packet, circulating round the ring, and fills its payload. Full packets intended for a particular destination are identified at their destinations and their payloads are copied in. Read packets are then released for use by other stations. Release of packets can be done by either the original sender (source release) or by the destination (destination release) [24, 25]. In this study we use the source release strategy.

Permission to send is given by special mini-packets called tokens. To maintain fairness, only one packet is sent when the token is held and then the token must be passed to the next station. Both the packets and the tokens travel in a single direction round the ring. Any station which receives a token, if it has a packet ready for transmission will send it, otherwise it passes the token to the next station on the ring.

As a model of the token ring, we assume the ring is symmetric and all stations have Poisson packet arrivals [26]. Hence we will look at only one station which is modelled as a queue with Poisson arrival and a single general type server, i.e. M/G/1. Since service requires the station to wait for a token, the service time is a function of the token rotation time T; this in turn depends on stations inserting packets in the ring. We will ignore the possibility of errors on the ring, and consider service complete when the packet is inserted into the ring.

Since every station during T will receive a token, to send a packet, then distribution of T can be calculated from the number of stations waiting to send packets enqueued. Knowing the distribution, we can then calculate the mean and the variance of the token time, which will be used in the M/G/1 queue model to calculate the mean queuing time.

For a Poisson packet arrival rate of λ packets (tokens) per second, and the mean waiting time $E(T)$, the offered traffic from each station (queue) or the probability that a station is in the queue is $a = \lambda E(T)$. Thus the probability that n out of N stations are in the queue is binomially distributed:

$$\text{prob (n queues busy)} = \binom{N}{n} a^n (1-a)^{N-n}$$

$$(7.78)$$

For such a binomial distribution, the mean and variance of the number of busy stations, which is equal to the number of packets on the ring:

$$E(n) = Na$$

$$(7.79)$$

$$V(n) = na(1-a)$$

$$(7.80)$$

Since each station exerts a fixed processing delay, δ, on each passing packet (token) irrespective of being active or idle, and also for each sent packet there is a propagation delay of, τ, then the relation between the mean token rotation time and mean number of packets in the queue is:

$$E(T) = E(n)\tau + N\delta = Na\tau + N\delta$$

(7.81)

Since the value of the processing delay δ is a constant, the variance is simply the variance of the number of stations scaled by τ^2:

$$V(T) = V(n)\tau^2 = Na(1-a)\tau^2$$

(7.82)

The offered traffic from each station, $a = \lambda E(T)$ can now be related to the mean number of active stations, by combining Equations 7.79 and 7.81. Therefore:

$$a = \lambda E(T)$$
$$\Rightarrow E(T) = \frac{a}{\lambda}$$
$$\Rightarrow \frac{a}{\lambda} = Na\tau + N\delta$$
$$\Rightarrow a = \frac{N\lambda\delta}{1 - N\lambda\tau}$$

(7.83)

Substituting a into Equations 7.81 and 7.82 the mean and variance of the token rotation time gives:

$$E(T) = \frac{a}{\lambda} = \frac{N\delta}{1 - N\lambda\tau}$$

$$V(T) = \frac{N^2\lambda\delta\tau^2}{(1 - N\lambda\tau)^2}[1 - N\lambda(\tau + \delta)]$$

(7.84)

In Chapter 4, it was shown that for a M/G/1 queue, the mean queuing time, $E(w_Q)$, is related to the second moment of the service time, $E(T^2)$,

$$E(w_Q) = \frac{\lambda E(T^2)}{2(1-a)}$$

(7.85)

Hence we need to calculate the second moment from the mean and variance;

$$E(T^2) = V(T) + \{E(T)\}^2$$

$$\Rightarrow E(T^2) = \frac{N^2\lambda\delta\tau^2}{(1-N\lambda\tau)^2}[1-N\lambda(\tau+\delta)] + \frac{N^2\delta^2}{(1-N\lambda\tau)^2}$$

(7.86)

This gives the mean queuing time for the token ring:

$$E(w_Q) = \frac{\lambda E(T^2)}{2(1-a)}$$

$$= \frac{\lambda}{2}\frac{(1-N\lambda\tau)}{[1-N\lambda(\tau+\delta)]}\left[\frac{N^2\lambda\delta\tau^2}{(1-N\lambda\tau)^2}[1-N\lambda(\tau+\delta)] + \frac{N^2\delta^2}{(1-N\lambda\tau)^2}\right]$$

$$= \frac{N^2\lambda^2\delta\tau^2}{2(1-N\lambda\tau)} + \frac{N^2\lambda\delta^2}{2[1-N\lambda(\tau+\delta)](1-N\lambda\tau)}$$

(7.87)

For the mean number of stations in the queue, Using the Little's formula, then

$$E(N_Q) = E(w_Q)\lambda = \frac{\lambda^2 E(x^2)}{2(1-a)}$$

$$E(N_Q) = \frac{N^2\lambda^3\delta\tau^2}{2(1-N\lambda\tau)} + \frac{N^2\lambda^2\delta^2}{2[1-N\lambda(\tau+\delta)](1-N\lambda\tau)}$$

(7.88)

Figure 7.29 shows the variation of $E(w_Q)$ with normalised load. The number of stations, N, is 50; the processing delay at each station, δ, is 6 bit times at 4 Mbit/s i.e. 1.5 μs; the packet transmission time, τ, is 128 μs or 1.024 ms corresponding to 64 or 512 byte packets (8-bit bytes), respectively, at 4 Mbit/s. The mean arrival rate is a proportion of 154.44 or 19.503s^{-1}, respectively, which corresponds to maximum normalised load at the two different packet lengths.

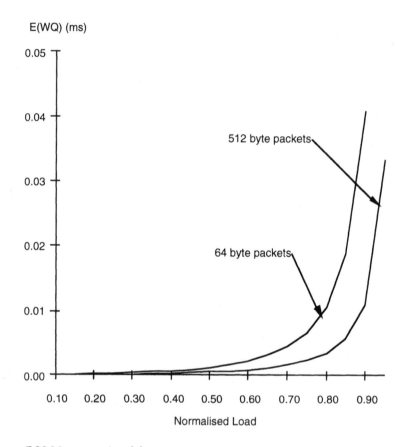

Figure 7.29 Mean queuing delay

7.13.2   Case 2: speech on ATM links

The asynchronous transfer mode (ATM) method of transmission combines the bandwidth flexibility of packet systems with the simplicity of circuit switched systems [10]. The (digitally coded) information from a source is assembled into fixed-size packets (cells) and when a cell is complete it is transmitted over the multiplex link in time division with other cells. In this way, low bandwidth services can be combined efficiently with high bandwidth services. Each cell contains a payload of **48** octets (**384** bits) and a header, which carries the address and some other information of **5** octets.

When we consider the transmission of speech on ATM links, it is not justifiable to consider the packets or cells to be generated at random in a Poisson process. As explained in Section 7.11, speech is produced in bursts ('talkspurts') interspersed with pauses. The resulting autocorrelation effects have a profound effect on the behaviour of the ATM multiplex system.

Let us suppose N speech channels are to be transmitted over the multiplex. The speech is encoded at the standard PCM rate of 64 kbit/s. Now unless the multiplex is operated at a speed of **64N** kbit/s or higher, which would defeat the transmission efficiency objective of ATM, there will be occasions when the instantaneous generation of cells from all the speech channels temporarily exceeds the capacity of the multiplex. The excess cells might be stored in a buffer, which would give rise to delay, but for the purpose of simplicity we will assume that there is no buffer at the input to the multiplexer. If then there is a temporary overload, cells must be discarded to the extent that the capacity is exceeded.

The payload of an ATM cell (**48** octets) will carry 6 ms of speech. We will further assume that the discard of a single cell will not normally be perceptible to the user if some fill-in technique is used, such as repeating the contents of previous cell. However, if two or more successive cells are lost, it is assumed that the deterioration in speech quality will be perceptible. We are required to find the maximum loading on the multiplex that will make the probability of the loss of two successive cells in a channel acceptably small.

Each channel may be modelled by the two-state Markov diagram shown in Figure 7.30, as explained in Section 7.11. The model is sampled at the multiplex cell transmission interval (**CTI**) and the transition coefficients have been adjusted accordingly as will be explained later.

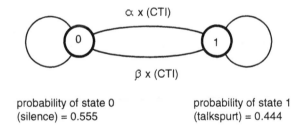

$$\alpha \times (CTI)$$

$$\beta \times (CTI)$$

probability of state 0
(silence) = 0.555

probability of state 1
(talkspurt) = 0.444

Figure 7.30 Single channel speech model

The corresponding multiplex model will then consist of states numbered 0 to N as shown in Figure 7.31. The state number is proportional to the number of cells generated during the CTI.

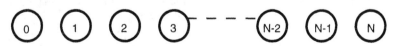

Figure 7.31 Multiplex speech model

The probability (π_s) that the multiplex model will be in state s is easily calculated from the binomial relationship

$$\pi_s = \binom{N}{s} p_1{}^s (1-p_1)^{N-s}$$

(7.89)

The state H corresponding to the speed of the multiplex link link is then fixed so that the loading on the multiplex is

$$L = 0.444 \frac{N}{H}$$

since the probability of a speech channel being in the active state is 0.444.

The cell transmission interval will depend on the speed of the multiplex. In this case, we have kept the rate at which the cells are generated by the N speech channels constant and adjusted the speed of the multiplex in accordance with the loading by setting the threshold state H. The transition coefficients for the single channel model will then depend on H, so that

$$\beta = 0.75\,dt = \frac{0.75 \times 384}{64000H}$$

While the model is in state $H + 1$, cells are discarded at the rate of $1/(H + 1)$; while the model is in the state $H + 2$, cells are discarded at the rate of $2/(H + 2)$ and so on. The total probability of a cell being discarded is then

$$p_u = \sum_{k=1}^{N-H} \frac{k}{H+k} \pi_{H+k}$$

(7.90)

To find the combined probability of a cell being discarded and then the subsequent cell in the same channel being discarded 6 ms later we proceed as follows. Consider a series of thresholds k (= 1, 2, 3, . . .) as shown in Figure 7.32. When the state of the multiplex model moves above a threshold, cells will be discarded as explained above and we are interested in the probability that the model will stay above a given threshold for 6 ms or longer.

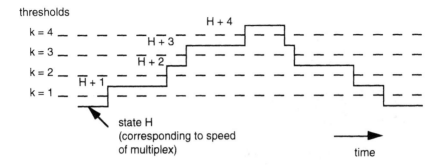

thresholds

state H
(corresponding to speed
of multiplex)

time

Figure 7.32 Thresholds in Markov model

Since the speed of the multiplex is **64 H** kbit/s and since the size of the ATM cell is **384** bits, the cell transmission interval is

$$\frac{384}{64H} = \frac{6}{H} \text{ milliseconds}$$

Hence the **6** ms interval corresponds to

$$\tau = \text{H CTIs}$$

For each threshold, we can now aggregate the states above and below the threshold, since we are mainly concerned with the behaviour of the system about the threshold. We will then have a series of two-state aggregate models corresponding to each threshold as shown in Figure 7.33

For each aggregate model, the probability that it will be in the state above the corresponding threshold, **k**, is the cumulative state probability for the multiplex model

$$F_k = \sum_{k=K}^{N-H} \pi_{H+k}$$

$$(7.91)$$

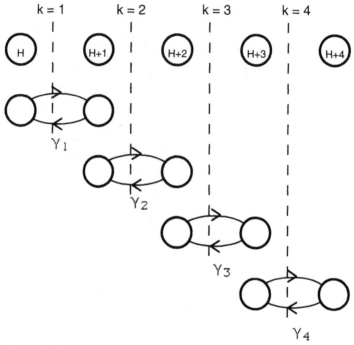

Figure 7.33   Aggregate model at threshold

Also for the multiplex model, the total probability 'flow' from states above the threshold to those below is given by

$$R = \sum_{j=1}^{N-H+k} \binom{H+k+j-1}{j} \beta^j (1-\beta)^{H+k-j} \pi_{H+k+j-1}$$

(7.92)

Then for the corresponding two-state aggregate model, the probability of transition from above the threshold to below is

$$\gamma_k = \frac{R}{F_k}$$

(7.93)

and the probability that the model will remain above the threshold for τ CTIs or longer is

$$\approx (1-\gamma_k)^{\tau} \qquad \text{if N is large}$$

$$(7.94)$$

Consider now the first threshold (k = 1). If the corresponding aggregate model is above the threshold at the first instant of sampling, one cell out of H+1 will be lost and if the model remains above the threshold for τ CTIs or longer, there is a 1/(H+1) chance that the loss will be in the same speech channel. Similarly if the model remains above the second threshold, the probability of initial cell loss will be 2/(H+2) and if it remains above threshold for τ CTIs or longer, there is 2/(H+2) chance that the loss will be in the same channel, and so on. However, if the model rises above the first threshold, it may also rise above the second threshold. Then to avoid double accounting, the second probability should be subtracted from the first.

Hence the total probability of two successive discards in the same channel interval τ is

$$P_c = \sum_{k=1}^{N-k} \frac{k^2}{(H+k)^2} [F_k(1-\gamma_k)^{\tau} - F_{k+1}(1-\gamma_{k+1})^{\tau}]$$

$$(7.95)$$

As an example, the probabilities of successive discards for a selection of multiplex loadings to carry 100 speech channels have been calculated and the results are shown in Figure 7.34. If we take the case of 71.7% loading, the state corresponding to the multiplex speed is state 62 and the CTI is 0.097 ms. The probability of successive cell losses in a channel is then 4.0×10^{-6}. The perceptible effect of cell loss will be felt on average about every 24 seconds.

This study illustrates the effects of autocorrelation in the coded speech. If the cells in each channel arrived at random as in a Poisson process, the probability of successive cell losses would then be simply $(1.43 \times 10^{-7})^2$ or 2×10^{-14}.

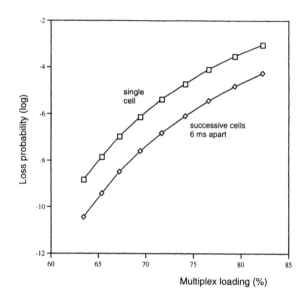

Figure 7. 35   Probabilities of cell losses in speech channels multiplexed over ATM links

7.14   References

1   PAPOULIS, A.: 'Probability, random variables and stochastic processes', McGraw Hill (1984)

2   CATTERMOLE, K. W. and O'REILLY, J. J.: 'Problems of randomness in communication engineering', Pentech Press (1984)

3   KLEINROCK, L.: 'Queueing sytems **1**: Theory' Wiley-Interscience, (1975)

4   McQUILLAN, J. M. and WALDEN, D. C.: 'The ARPA network design decisions', Computer Networks, **1**, (1977) pp. 243–289

5   MEDCRAFT, D. W. F.: 'Development of the UK packet switched services', *Proceeding Networks*, **80**, London (1980)

6   SCHWARTZ, M.: 'Telecommunication networks, protocols, modelling and analysis', Addison Wesley (1987)

7   DEASINGTON, R. J.: 'X.25 explained: protocols for packet switching networks', 2nd edition, Addison-Wesley (1988)

8   ITU-T Recommendation X.25: 'Interface between data terminal equipment (DTE) and data-circuit terminating equipment (DCE) for terminals operating in the packet mode and connected to public data networks by dedicated circuits'

9   ITU-T Recommendation I.121, 'On the broadband aspects of ISDN', Blue Book, Geneva, Switzerland, (1989)

10   CUTHBERT, L. G. AND SAPANEL, J-C.: 'ATM - The broadband telecommunications solution', The Institution of Electrical Engineers, (1993)

11   de PRYKER, M.: 'Asynchronous transfer mode, solution for Broadband ISDN', 3rd edition, Ellis Horwood, (1995)

12   AHMADI, H. and DENZEL, W.: 'A survey of modern high-performance switching techniques', *IEEE Journal on selected Areas in Communications*, **7**, no.7, September (1989)

13   COPPER, R. B.: 'Introduction to queuing theory', North Holland, (1981)

14   BULLINGTON, K. and FRASER, J. M.: 'Engineering aspects of TASI', Bell System Technical Journal, **38**, (1959) pp.353–364

15   BRADY, P. T.: 'A statistical analysis of on-off patterns in 16 conversations', *Bell System Technical Journal*, **47**, (1968) pp.73–91

16   GRUBER, J.: 'A comparison of measured and calculated speech temporal parameters', *IEEE Trans. on Comm.* **COM-30**, (1982) pp.728–738

17   LIOU, M.: 'Overview of px64 kbit/s video coding standard', *Communications of the ACM*, **34**, no. 4, (1991) pp. 59–63

18   Le GALL, D.: 'MPEG: A video compression standard for multimedia applications', *Communications of the ACM*, **34**, no. 4, (1991) pp.305–313

19   HUANG, S-S.: 'Source modelling for packet video', *Proc.ICC '88* (1988) pp.1262–1267

20   HUGHES, C. J. , GHANBARI, M, PEARSON, D. E., SEFERIDIS, V. and XIONG, J. 'Modelling and subjective assessment of cell discard in ATM video', *IEEE Trans. on Image Processing*, **2**, (1993) pp.212–222

21   GHANBARI, M.: 'Two-layer coding of video signals for VBR networks', *IEEE Journal on Selected Areas in Communications*, **7**, (1989) pp.771–781

22   MAGLARIS, B. *et al* :'Performance models of statistical multiplexing in packet video communications', *IEEE Trans. on Comm.* **COM-36**, (1988) pp.834–844

23   SEN, P.: 'Models for packet switching of variable bit-rate video sources', IEEE J. Select Areas in Comm. **7**, (1989) pp. 865–869

24   FALCONER, R. M, ADAMS, J. L. and WALLEY, G. M.: 'A simulation study of the Cambridge ring with voice traffic', *British Telecom Technology Journal*, **3**, no. 2, (1985) pp. 85–91

25   FALCONER, R. M. and ADAMS, J. L.: 'Orwell: A protocol for an Integrated services local networks', *British Telecom Technology Journal*, **3**, no. 4, (1985) pp. 27–34

26   MOLLOY, M.K.: 'Fundamentals of performance modelling', MacMillan (1989)

Chapter 8

Introduction to reliability

8.1 Reliability performance

The estimation of the reliability of a system is an important branch of performance engineering. As in most cases of attempting to predict the future, we can do so only in probabilistic terms. Just as in teletraffic studies, the precise behaviour of the system is dependent upon the behaviour of the individual users, so in estimating the reliability performance we are at the mercy of more or less random failures of the individual components. However, even though we cannot predict the precise time of failure of each component, it is useful to know the probabilities and consequences of component failure. The performance of the system is then given in terms of the probabilities of complete or partial failure and the resulting economic and other consequences may be estimated. Managerial judgement is required to decide then whether or not the reliability of the system is acceptable but at least decisions can be made in the light of the calculated probabilities.

In cases where human life or safety is concerned, reliability design can be of supreme importance. Two-engined aircraft now fly across the Atlantic, whereas until recently it was thought that only three or four engines provided an acceptable standard of reliability. The reliability of the jet engines used has been calculated and it has been decided that the probability of both engines failing during the transatlantic flight represents an acceptable risk.

8.1.1 Cost effectiveness

Apart from the human safety aspects, which in some circles cannot be valued in monetary terms, the cost effectiveness of reliability can be estimated. First, a cost can be imputed to system failure, although a degree of judgement may be required. Direct costs such as the cost of repair may be estimated with fair accuracy but even here there are some uncertainties. The cost of sending a technican to repair the fault or the cost of transporting the equipment to a central repair depot needs to be taken into account. This needs to be considered even for a simple system such as a TV receiver. An extreme case is that of submarine cable repeaters, where the cost of the replacement repeater is much less than the cost of sending a cable ship to recover the faulty unit.

Another item on the debit side is the cost of the system being unavailable until the repair is completed. The loss of revenue arising from breakdown in an electricity supply or a telecommunications network may be relatively small but the indirect costs of loss of factory or office output will be much greater. Even if the electricity supply or telecommunications company does not have to bear such costs, unless agreements with the customers contain penalty clauses, the loss of goodwill is effectively a cost penalty.

It is usually possible to improve the reliability of a system – at a cost. There are many ways of achieving a higher standard of reliability including:

- better design: some faults are due to a failure to design the system properly rather than to component breakdowns. This is particularly true in the case of software systems, where so-called faults are really design failures;

- more reliable components: components may be manufactured to a more stringent specification;

- derating components: generally components that are run well inside their rated heat dissipation will have a longer life;

- wider operating tolerances: components do not always break down completely and a shift of one or more characteristics outside the operating range may lead to system failure. This is an important factor to be considered at the design stage. In fact, some high speed digital systems would not work at all if all the logic gate delays were on the limits of their stated tolerances;

- 'burn in' to eliminate early failures: some components may pass the acceptance tests but break down after only a few hours' operation. The high cost of replacing the component in the field may be avoided if the system is initially run for some time under factory conditions;

- power supply protection: the lifetimes of active components may be reduced if subjected to over-voltage power supplies, including transients;

- improved maintenance procedures: in cases where the lifetime of a component can be determined within relatively close limits, it may be worthwhile to institute a programme of regular replacement; more importantly, care should be taken to ensure that maintenance procedures are carried out in such a way that intrusion of the maintenance technician does not of itself introduce further faults;

- redundancy of components, subsystems or network: introducing some form of redundancy may be the only way to improve reliability to the degree required after some of the measures given above have been applied. However, the redundant units are themselves subject to possible failure as are changeover mechanisms;

- fault tolerant software: systems which are partly or wholly controlled by software may be designed so that both software and hardware faults can be recognised and the system reconfigured to work round the faulty unit;

In theory at any rate, if the direct and indirect costs of system failure can be calculated, it is possible to estimate the cost effectiveness of a reliability design target as shown in Figure 8.1. There is an optimum point but, since in most cases the cost rises relatively slowly with improved reliability, it is usually advisable to aim for a higher standard of reliability than that given by the minimum total cost.

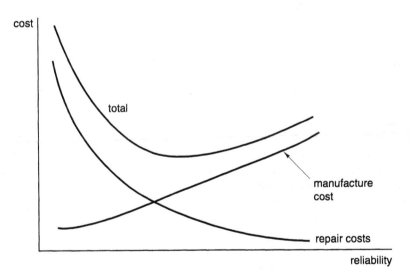

Figure 8.1 Cost of reliability

8.2   Reliability and availability

Although 'reliability' is used in a general sense, it can be given a precise meaning as the probability that a system will continue to perform satisfactorily for a given period of time (t). Some systems may be designed on the basis that they will be scrapped if they fail. Examples include desk calculators, telephone instruments, lighting units, vehicle tyres etc. In this case, we are only concerned with the probability of transition from the '0' state (operational) to the '1' state (failed) as illustrated in Figure 8.2a.

A component or system is said to be in a failed state if it is unable to perform the function for which it was intended. This applies to components that drift outside their prescribed operating limits as well as to complete breakdowns.

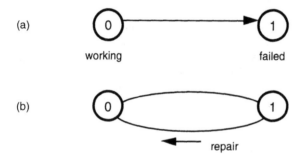

Figure 8.2 Transitions between working and failed states

It is important for a manufacturer to assess reliability since early failures will often result in the systems having to be replaced under warranty. Apart from the loss of consumer confidence, this can be an expensive process. It may well be cost effective to spend more in the factory to improve reliability than to replace items in the field.

In other cases, the system is repaired and put back into service as soon as possible. We are then concerned with the probability of a reverse transition from the failed state to the working state as well as with the forward transition as shown in Figure 8.2b. In this case the important parameter is the availability which depends on both the reliability and the repair time.

The following definitions apply (see Figure 8.3).

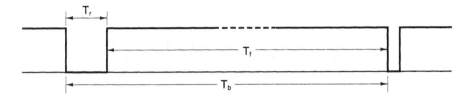

Figure 8.3 Failure and repair definitions

Mean time to failure (MTTF) – the average time between the completion of a repair and the next failure, $E[T_f]$

Mean time to repair (MTTR) – the average time between the occurrence of a failure and the restoration of the system to service, $E[T_r]$

Mean time between failures (MTBF) – the average time between successive failures, $E[T_b]$

Failure rate – the average rate at which breakdowns occur; this is not necessarily constant and may vary with time. In some publications, hazard rate is used for the breakdown rate under non-repairable conditions and failure rate when repair is intended but the distinction is by no means universal. Failure rates may be measured in faults per 10^6 h, per cent per 1000 h, or even in faults per year. In electronic systems, it is often convenient to use faults per 10^9 h, known as 'FITS'.

In large complex systems, operation may be possible even when several faults are present. The behaviour may then be represented by a Markov chain as shown in Figure 8.4. The analogy with a Poisson process to describe call or packet arrivals in teletraffic studies will be evident.

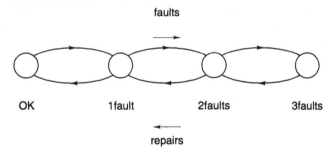

Figure 8.4 Markov representation of fault and repair process

8.3   Limitations of reliability estimation

Reliability estimation, as in any other form of prediction, cannot be expected to give exact answers. A prediction cannot be better than the data used and, even then, the reliability or availability can only be estimated in terms of probabilities. Nevertheless, reliability calculations are of considerable importance in system design because of the need to achieve a satisfactory standard of reliability as seen by the users. The need for economical engineering is an additional reason why reliability calculations are vital; unreliability can be very expensive. In some cases, different design approaches are possible and it is often useful to compare the alternative designs in terms of reliability. Furthermore, reliability calculations enable the weakest parts of a system to be identified. If a higher level of reliability is required, then improvements should preferably be made in the design of the least reliable components or subsystems.

8.4   Basic relationships

In most cases of performance engineering, we are concerned with random events that occur within a continuum of time. Alhough the events occur at random, the mean rate of occurrence is usually assumed to be constant. Where reliability studies are concerned, this is not always so and we sometimes need to consider cases in which the rate of occurrence of events is accelerating or decelerating. The theory is therefore developed to meet the general case and the constant event rate situation is then shown to be a simplification.

8.4.1   Reliability functions

Suppose a system operates satisfactorily for a time X and then fails. Then X is known as the lifetime of the system. The reliability may then be defined as the probability that a system will perform satisfactorily for a given period of time, that is

$$R(t) = P[X > t]$$

Consider the probability of failure between time and $t + \delta t$. The probability distribution function, $f(t)$ for such an event is defined by

$$f(t)\delta t = P[t \leq X \leq (t + \delta t)]$$

The corresponding cumulative distribution function $F(t)$, that is the probability that the failure takes place up to time t, is

$$F(t) = P[X \le t]$$

Then the reliability

$$R(t) = P[X > t]$$

$$= 1 - F(t)$$

Hence

$$R(t) = 1 - \int_0^t f(t') \, dt'$$

(8.1)

or

$$f(t) = -\frac{d}{dt} R(t)$$

(8.2)

Table 8.1 illustrates the failure and reliability functions by means of a somewhat artificial example of 1000 units that were placed on test at day zero. The number of units that failed by the end of each day was noted and the functions calculated.

8.4.2 Failure rate

Failure rate is understood in general terms as the average number of failures relative to the population, divided by the interval in which they occur. It may be given a more precise definition by supposing that the system performs satisfactorily up to time $X = t$ and then fails at $X < t + \delta t$. Then the failure rate ($\lambda(t)$) is defined as the conditional probability

$$\lambda(t) \, \delta t = P[X < t + \delta t \mid X > t]$$

$$= \frac{P[(X > t) \, \& \, (X < t + \delta t)]}{P(X > t)}$$

Table 8.1   Example of reliability test

day	failures during day	prob. density f(t)	total failures	failure CDF F(t)	reliability R(t)	mean no. at risk during day	failure rate $\lambda(t)$
1	75	.075	75	.075	.925	962.5	.0779
2	50	.050	125	.125	.875	900	.0556
3	34	.034	159	.159	.841	858	.0396
4	23	.023	182	.182	.818	829.5	.0277
5	16	.016	198	.198	.802	810	.0198
6	12	.012	210	.210	.790	796	.0151
7	10	.010	220	.220	.780	785	.0127
8	8	.008	228	.228	.772	776	.0103
9	8	.008	236	.236	.764	768	.0104
10	7	.007	243	.243	.757	760.5	.0092
11	7	.007	250	.250	.750	735.5	.0093
12	7	.007	257	.257	.743	746.5	.0094

It follows that

$$\lambda(t)\,\delta t \;=\; \frac{f(t)\,\delta t}{R(t)}$$

or

$$\lambda(t) \;=\; \frac{f(t)}{R(t)} \;=\; -\,\frac{1}{R(t)}\,\frac{d\,R(t)}{dt}$$

$$(8.3)$$

The failure rate is sometimes referred to as the instantaneous failure rate or the hazard rate.

It is important to distinguish between the failure probability density and the failure rate. Referring back to Table 8.1, it will be seen that on the 6th day for example, there were 12 failures so that, since we started with 1000 units on test, the probability density was 0.012. However, at the end of the 5th day (beginning of the 6th day) there were $1000 - 198 = 802$ units still running and at the end of the 6th day there were $1000 - 210 = 790$ active units. The mean population of active units during the 6th day was therefore 796 units and the failure rate calculated relative to this population was $12/796 = 0.0151$.

We may also derive some other relationships between the reliability functions and the failure rate. Substituting from Equation 8.2 and integrating Equation 8.3 gives

$$\int_0^t \lambda\,(t')\,dt' \; = \; - \; \ln\,[\,R(t)\,]$$

hence

$$R(t) \; = \; \exp\left\{-\int_0^t \lambda\,(t')\,dt'\right\}$$

(8.4)

and the probability density function

$$f(t) \; = \; \lambda\,(t)\,\exp\left\{-\int_0^t \lambda\,(t')\,dt'\right\}$$

(8.5)

8.4.3 Mean time to failure

The mean time to failure (see Figure 8.3) is defined as the expected value $E[t_f]$ of the failure time, t_f.

Then the MTTF is

$$M(t) \; = \; \int_0^\infty t\,f(t)\,dt$$

(8.6)

Here again, the MTTF is defined in a general way as a function of time. This is often necessary if we wish to consider the reliability and failure rate over the complete life of a unit.

8.4.4   Failure conditions

Many electronic systems suffer high failure rate during the early part of their life (the 'burn-in' period) as shown in Figure 8.5a. For example, failures may be caused by misalignment of the masks used for printed wiring boards or semiconductor devices, or by poor soldered connections. High resistance spots cause local heating and this leads to breakdown after a short period of use. In mechanical systems, poor materials or badly toleranced machining may lead to early failure. Software shows a similar characteristic although for a different reason. In this case, the failure rate decreases as 'bugs' (that is, design faults) are cleared.

Similarly, failure rates tend to increase towards the end of the life cycle due to various wear out mechanisms (Figure 8.5b). In electronic systems, the wear out process may take the form of water ingress into so-called hermetically sealed semiconductor device packages. Nowadays, this effect may be delayed for several years. For software systems, there may be no wear-out period if virtually all the design faults are cleared but in practice an increase in failure rate may occur due to increased loading on the system and the consequences of the people responsible for maintenance becoming more distanced from the original design team.

The total life cycle failure rate characteristic for electronic systems consists of three parts, namely the burn-in and wear-out failure rate-out periods, and the time in between of virtually constant failure rate as shown in Figure 8.5c. The resulting characteristic is often referred to as a 'bathtub' curve. In many cases, the constant failure rate condition lasts for tens of years. For mechanical systems, the wear-out process starts as soon as the equipment is operated although failure, usually caused by a component part wearing beyond its tolerance limit, may take some time.

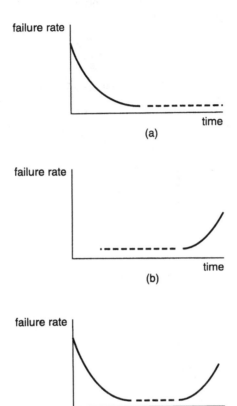

Figure 8.5   Failure rate conditions

8.5   Reliability modelling

In order to make predictions of the reliabilities of systems, it is useful if we can use simple mathematical functions to model either the failure rate or one of the reliability functions. The other functions may then be derived using the relationships given above.

8.5.1   *Constant failure rate*

If the system is operating at the bottom of the bathtub curve shown in Figure 8.5c the failure rate is assumed to be constant

$$\lambda(t) = \text{a constant, } \lambda$$

This implies a 'memoryless' property. Failures occur at random (Poisson process) as do calls in a telephone system.

From Equation 8.5, the PDF,

$$f(t) = \lambda(t) \exp\left\{-\int_0^t \lambda(t') \, dt'\right\}$$
$$= \lambda e^{-\lambda t}$$

$$(8.7)$$

and from Equation 8.5, the reliability becomes

$$R(t) = e^{-\lambda t}$$

$$(8.8)$$

These functions are plotted in Figure 8.6a.

The mean time to failure is then simply

$$\text{MTTF} = \frac{1}{\lambda}$$

$$(8.9)$$

8.5.2   *Lifetime PDF having a normal distribution*

In mechanical systems such as bearings, vehicle tyres, relays, etc. the expected life may be estimated with fair accuracy. If the rate of use is not constant, it may be more appropriate to use the number of operations rather than time as the independent variable. The mean of the lifetime probability distribution function, $f(t)$ is the predicted life and failure events may be expected to occur randomly at times on either side of the mean. Under these conditions, the reliability performance may usually be modelled by assuming a normal distribution.

The PDF takes the form

$$f(r) = \frac{1}{\sigma \sqrt{2\pi}} \exp\left[-\frac{(t-t_0)^2}{2\sigma^2}\right]$$

(8.10)

and from Equation 8.1, the reliability is given by

$$R(t) = 1 - \frac{1}{\sigma\sqrt{2\pi}} \int_0^t \exp\left[-\frac{(t-t_0)^2}{2\sigma^2}\right] dt$$

(8.11)

where σ is the standard deviation.

The failure rate may then be found from Equation 8.3

$$\lambda(t) = \frac{f(t)}{R(t)}$$

In this case, we have started by assuming a model for the lifetime PDF instead of the failure rate because of the lack of an explicit integral form for the normal distribution. The functions are plotted in Figure 8.6b.

8.5.3   Weibull distribution

This is a very flexible distribution and may be used to model a wide range of failure behaviour simply by changing the parameters. In this case we start from the failure rate, $\lambda(t)$ and assume it follows a power law

$$\lambda(t) = \frac{m}{\theta}(t/\theta)^{m-1}$$

(8.12)

where θ is a scale parameter and m is a shape parameter.

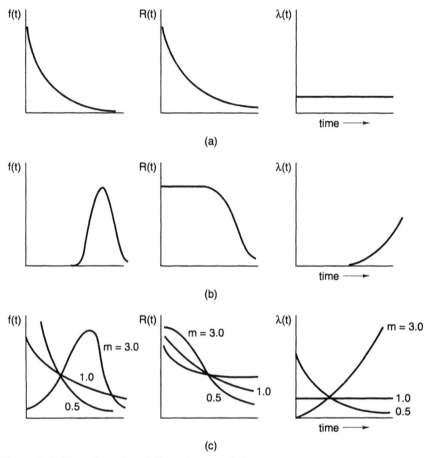

Figure 8.6   Time dependent failure characteristics

  (a)   constant failure rate
  (b)   log–normal distribution
  (c)   Weibull distribution

If m < 1, the failure rate decreases with time whereas if m < 1 the failure rate is an increasing function. The special case of m = 1 corresponds to a constant failure rate. Thus any part of the bathtub curve shown in Figure 8.6c may be modelled by choosing a suitable value for m.

From the failure rate, the PDF may be found using Equation 8.3

$$f(t) = \frac{m}{\theta} (t/\theta)^{m-1} \exp\left(-(t/\theta)^m\right)$$

(8.13)

This equation may appear to be clumsy to use in practice but the cumulative distribution function given by integration has a simple exponential form

$$F(t) = 1 - \exp\left[-(t/\theta)^m\right]$$

(8.14)

and the reliability

$$R(t) = \exp\left[-(t/\theta)^m\right]$$

(8.15)

These relationships are plotted in Figure 8.6c.

The mean of the Weibull distribution may be shown to be

$$\mu = \theta\, \Gamma\!\left(1 + \frac{1}{m}\right)$$

and the variance

$$\sigma^2 = \theta^2 \left\{ \Gamma(1 + 2/m) - \left[\Gamma(1 + 1/m)\right]^2 \right\}$$

where the gamma function is defined by

$$\Gamma(\alpha) = \int_0^\infty t^{\alpha - 1} \exp(-t)\, dt$$

A third parameter is sometimes used in the Weibull distribution to shift the zero of the time scale. This is achieved simply by writing $t - t_0$ in the above expressions (for $t \geq t_0$) and is useful in modelling systems in which a threshold amount of wear takes place before any failures occur.

8.6 Components

The reliability of most electronic systems depends on the reliability of the individual components. 'Components' in this sense mean not only devices such as capacitors, transistors etc. but also metallic connections such as tracks on printed circuit boards, and plug and socket connectors. Although connections are usually regarded as very reliable if the manufacturing process is carefully controlled, they are much more numerous than other components and so may make a significant contribution to the system unreliability.

Improvements in semiconductor technology and an increasing scale of integration have led to steadily improving reliability of microelectronic components. Low failure rates have become essential because of the increasing scale of integration. In the early days of integrated circuits containing one or two 'gates', failure rates of 1000 FITS were not uncommon. (A FIT is one failure in 10^9 h). This was quite acceptable since it corresponded to a MTTF of over 100 years. A modern LSI chip often contains the equivalent of 10 000 or more gates and might have a similar failure rate. The failure rate per gate has improved by four orders of magnitude. In some cases, the components have become so reliable that it becomes difficult to estimate the failure rate for system reliability calculations.

Packaging, particularly for active components, is an important factor in determining the reliability. Early packages were designed to make reliable connections and to dissipate heat. As the scale of integration has increased, heat dissipation has increased but the trend has been counteracted by newer technologies. The main cause for failure is usually the ingress of moisture and ceramic packages are generally more reliable than plastic. In hybrid packaging technology, microelectronic chips and other components are mounted on a ceramic or other substrate and the whole assembly is then sealed under a plastic coating. In this case, the 'component' is a complete sub-assembly. A high degree of reliability is possibly but only if the manufacturing process is carefully controlled.

8.7   Estimating component reliability

The basic method of estimating the reliability of a particular type of component or subsystem is to place a large number of the components on test under controlled conditions and note the times at which failures occur. In the case of mechanical components, the number of operations up to each failure is noted. Raw data gathered in this way can give useful design information but the preferred approach is to attempt to fit the data to one of the distributions mentioned in the previous section. The parameters of the distribution can then be determined. Statistical techniques and computer programs exist for fitting experimental data to mathematical distributions but a simple graphical approach is often perferred since it is possible to see at a glance the goodness of fit.

8.7.1   Exponential distribution (constant failure rate)

From a knowledge of the failure mechanism or from earlier experience with similar devices, it may be possible to assume a constant failure rate. By determining the constant λ from a controlled experiment, the behaviour of the device over its life in the field may be estimated.

From the expression for reliability (Equation 8.8)

$$R(t) = e^{-\lambda t}$$

by taking logarithms of both sides of the equation we have

$$\log\left[\frac{1}{R(t)}\right] = \log\left[\frac{1}{1-F(t)}\right]$$

$$= \log_{10}(e)\ \lambda t$$

$$= 0.4343\ \lambda t$$

It is usual to work in terms of the unreliability function $F(t)$ rather than $R(t)$. Since the testing involves noting the times of device failure, $F(t)$, may be observed directly. If out of a total batch of N components, i failures are observed by time t, the estimate of the failure CDF is

$$\text{est}\ [F(t)\ ] = \frac{i}{N}$$

In some cases, particularly where the test is continued until all the units have failed, a better estimate of the cumulative failure distribution is

$$\text{est}\ [\ F(t)\ ] = \frac{i}{N + 1}$$

since with a larger sample, some of the units might still be working at the time the original test had to be discontinued. If the testing is stopped arbitrarily before all the units have failed, as in this case, the data is referred to as 'censored'.

For example, suppose a batch of **100** microelectronics components were tested and the results obtained were as shown in Table 8.2.

Then by plotting $\log[1/(1 - F)]$ against time t (or alternatively using log-linear graph paper) a straight line should result. Pronounced departure from a straight line relationship would indicate that the original assumption of a constant failure rate was incorrect. The resultant graph is shown in Figure 8.7.

The slope of the graph is **0.018 / 104** and therefore the failure rate is

$$\frac{0.018}{10^4}\ \ln(10) = 4.1 \times 10^{-6}$$

Of course, with 8% failures in about **20 000** hs, it might be expected that the failure rate would be about 4×10^{-6} but the plot was necessary to validate the assumption of constant failure rate.

Table 8.2   Test of components having constant failure rate

failure number (i)	time to failure (hours)	est. failure distribution function (F)	log[1/(1-F)]
1	2235	0.0099	0.00432
2	4980	0.0198	0.00869
3	7448	0.0297	0.0131
4	9552	0.0396	0.0175
5	12056	0.0495	0.0221
6	15002	0.0594	0.0266
7	17124	0.0693	0.0312
8	19984	0.0792	0.0358

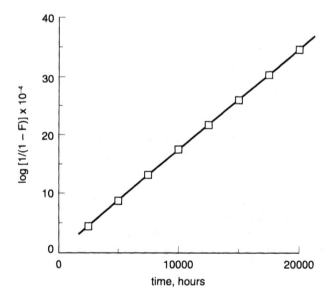

Figure 8.7 Plot for results given in Table 8.2

8.7.2   Weibull distribution

If the form of the reliability distribution is not known, it is usually safer to assume a Weibull distribution. If it is found that the value of the constant m as estimated from the test data is nearly equal to unity, then it may be assumed that the exponential distribution is acceptable.

The Weibull distribution, Equation 8.15

$$R(t) = \exp\left[-(t/m)^m\right]$$

may be rewritten in logarithmic form and in terms of the cumulative distribution function, $F(t)$

$$\ln\left[1/(1-F)\right] = (t/\theta)^m$$

Then taking the logarithms of both sides of the equation

$$\log\left[\ln\left(1/(1-F)\right)\right] = m\log t - m\log\theta$$

(8.16)

This is of the form $y = mx + c$ and if the data are plotted in this form, a straight line should result provided that it approximates to a Weibull distribution. The slope of the graph gives the shape parameter, 'm'. The scale parameter θ may be found by noting that when the left hand side of equation is zero, then $\theta = t$. For example, a batch of 23 sub-assemblies were put on test and failures were observed as follows. (The estimate of the failure distribution is of course $F(t) = i/N$)

Table 8.3   test of components having Weibull failure rate

failure number (i)	time to failure (hours)	est. failure cum. function (F)	log (1/(1-F))
1	28	0.04	−1.39
2	83	0.08	−1.08
3	154	0.12	−0.89
4	342	0.16	−0.76
5	469	0.20	−0.65
6	558	0.24	−0.48
7	907	0.28	−0.41

The results of the tests are plotted in Figure 8.8. It is of course possible to use log-log graph paper instead of taking logarithms as in Equation 8.16. For more serious testing, special Weibull probability paper is used. Then from Equation 8.16, it will be seen that the slope gives the constant, m in the Weibull distribution. Note that the actual slope (as measured directly on the graph paper) should be used. In this case,

$$m = \frac{1.7}{2.55} = 0.67$$

and when the ordinate is equal to zero, then

$$\log(t) = 3.55$$

therefore θ =   antilog(3.55) = 3548

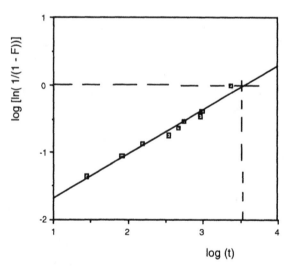

Figure 8.8   Plot of results given in Table 8.3

Since the value of the shape parameter, m is less than unity, the failure rate is decreasing with time. This represents the burn-in period.

Because of the flexible nature of the Weibull distribution, it can be used to represent almost any failure characteristic by suitable choice of shape and scale parameters and a straight line will normally result if the plot is made as shown above. A significant departure from the straight line relationship normally indicates that the zero point in time is incorrect and the plot should be made in terms of $(t - t_0)$ rather than t.

8.8 Accelerated life testing

The problem with newly designed components is that it is often necessary to put them into service before the reliability characteristics can be determined. In some cases, the reliability can be estimated by analogy with similar components but this carries a risk in that minor changes in the manufacturing process may significantly alter the reliability. The alternative is to adopt some form of accelerated life testing.

For mechanical components, the solution is to test the component by increasing the rate of operation. A switch for example that is intended to be used one or two times a day may be operated several times a minute under test conditions. The failure rate after several years of normal life may then be estimated from the results of a few hours' testing.

For microelectronics components manufactured under carefully controlled conditions, it has been found that the main cause of failure is diffusion of semiconductor materials or of water ingress into the package. It is thus similar to many chemical processes that obey the Arrhenius model [1] relating the speed of the reaction to the temperature:

$$v_r = C \exp\left(-\frac{A}{T}\right)$$

(8.17)

where A is a constant related to the activation energy for the process
and T is the absolute temperature of the junction in K.

The temperature at which the test is carried out is only limited by the type of package used and testing is often carried out at temperatures of 150 °C (423 K) or higher. The actual junction temperatures have to be estimated from the currents invoved and the layout on the substrate. In this way, the ageing process relative to operation at normal temperature may be increased by up to about three orders of magnitude.

Because of the importance of being able to estimate the reliability of a new electronic component as soon as possible after the prototype has been made, the art of accelerated testing has been refined considerably and techniques such as step stress testing have been developed. The step stress method involves subjecting the component to operation at a carefully regulated series of temperature steps and the theory has been developed so that the reliability at a normal working temperature may be calculated.

8.9   Confidence intervals

In Chapter 3, we introduced the topic of confidence intervals in relation to simulation techniques. In reliability testing just as in simulation, it is useful to know to what extent the results can be relied on. In both cases, we are considering random samples, so that the results obtained are not abolute and the degree of accuracy is best expressed in terms of probabilities such as confidence intervals In most cases we assume that any random errors follow a normal distribution if the number in the sample is sufficiently large. Suppose a sample of N values has a mean m and standard deviation s. Then the standard deviation of the sample means or the standard error of the estimate is

$$s' = \frac{s}{\sqrt{N}}$$

On the basis of normally distributed errors, the population mean can be said with 90% confidence to lie within the range $\pm 1.65 s'$ (95.5% confidence within $\pm 2s'$).

A case of particular interest in reliability testing is that in which no failures arise during the testing period. It would be extremely rash to assume that the component was 100% reliable but even from such a negative test it is possible to obtain useful reliability information. Suppose N devices are tested and that p is the (unknown) probability of failure. Then if we took a large number of such samples, the number of failures per sample would follow a Binomial distribution and the probabiltiy of r failures in a sample

$$q_r = \binom{N}{r} p^r (1-p)^{N-r}$$

$$(8.18)$$

and the probability of zero failures in a sample

$$q_0 = (1-p)^N$$

$$(8.19)$$

Since our particular sample gave zero failures, we can be $1 - q_0$ confident that the expected number of failures per sample was Np or less. Note that for a higher value of p, the probability of zero failures in the sample decreases, so that if zero failures are observed we can be more confident that p is below the limit. Alternatively if we require 90% confidence $(1 - q_0 \leq 0.9)$ then

$$p \leq 1 - (0.1)^{1/N}$$

For example, if a test of 50 items gives zero failures, we can be 90% confident that the failure probability was less than 0.045 or, in other words, the reliability at the time the test was ended was 95.5%.

The theory has been developed so that the confidence limits for any number of failures may be calculated. The reader is referred to specialist books on reliability for further information (see References).

8.10 Reliability standards

Bodies such as the US Department of Defense (DoD) [2] and the British Standards Institution [3] produce standards for components manufactured to an agreed specification. These give failure rates for a wide variety of electronic components operating under a range of operating conditions. The standards are widely used by both component and system manufacturers. Testing of components to meet the required standards does, of course, involve additional expense and cheaper components are often used for 'consumer' electronic equipment.

8.11 Burn-in

If the initial failure rate is one that decreases with time, then it may be worth while to operate the components under controlled conditions for a period of time before assembly. The process is known as 'burn-in'. Additional expense is involved in both the cost of the burn-in facilities and in the additional capital required for equipment which has been made but cannot be sold until after the burn-in period. On the other hand, it is often possible to reduce the cost of expensive field servicing during the warranty period.

The effect of burn-in is shown in Figure 8.9. Suppose the unit is required to operate for time t_0. The reliability under normal conditions is $R(t_0)$. Now let the burn-in time be t_b. The conditional probability that the unit will survive for $t_b + t_0$ given that it has survived the burn-in is then

$$R[(t_b - t_0) \mid t_0] = \frac{R(t_b + t_0)}{R(t_0)}$$

$$(8.20)$$

If the failure rate is constant (Poisson process) the reliability remains constant irrespective of the time at which the measurement is started. There is then no advantage in a burn-in procedure. However, electronic components sometimes work perfectly satisfactorily for a few hours and then fail. This can be due to a variety of causes, as mentioned in Section 8.4.3. By eliminating those which have undetected manufacturing defects that make them liable to early failure, the reliability of the remaining components will be enhanced.

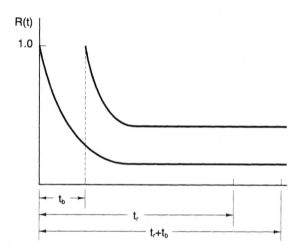

Figure 8.9   Improvement in reliability as a result of burn-in

8.12   System reliability assessment

In the first part of this section, we consider the reliability of systems without reference to any repair action. In systems such as some consumer electronics and missiles, it is not intended that a repair should be carried out. In other cases, such as aircraft control systems, the effects of failure would be so disastrous that a failure might result in physical destruction of the system. The objective is then to design the system so that the probability of failure is so low that it constitutes an acceptable risk. In other systems, repairs are possible and the objective is usually to design the system with an acceptable down-time performance.

8.12.1 Multi-component systems

The easiest type of system to analyse in reliability terms is one in which component failures are independent of one another and the system fails if one component fails. In this case

$$R(t) = P(X > t)$$

$$= P(x_1 > t \ \& \ x_2 > t \ \& \ .....\& \ x_n > t)$$

$$= R_1(t) . R_2(t) ....R_n(t)$$

In general terms,

$$R(t) = \prod_{j=1}^{n} R_j(t)$$

$$(8.21)$$

For a constant failure rate,

$$R(t) = \exp[-(\lambda t)]$$

$$= \exp[-(\lambda_1 + \lambda_2 + ... + \lambda_n)t]$$

Hence

$$\lambda = \sum_{j=1}^{n} \lambda_j$$

$$(8.22)$$

That is, the failure rate for the system is the sum of the failure rates of the individual components, which is an obvious result.

The calculation of the reliability of large systems possibly involving tens of thousands of components is a straightforward but tedious operation. However, in computer aided design (CAD) systems, reliability calculations are often carried out automatically as an integral part of the design process. The failure rate or MTTF of each type of component is stored in the system and the reliability information is associated with each item in the component lists. In addition to normally identifiable components, it is usual to include tracks on printed circuit boards and plug and socket connections. The failure rate for a subassembly is calculated and automatically updated when any component changes are made.

8.12.2 Maintenance spares

In large systems, spare components or subsystems are normally held close to the equipment for quick replacement in case of failure. When a unit is replaced, a new one is usually ordered or the subsystem repaired but this may take some considerable time. The level of maintenance stocks is an important issue in the system operation. Excessive provision of spare units will provide good security of service but will be expensive. If no spare unit is available in the case of failure, the system may be out of service for a long period. Even if redundancy has been built into the system, any faults should be rectified promptly to reduce the risk of catastrophic breakdown.

Let c be the number of components or subsystems of a particular type in the system and let λ be the failure rate for each component (assumed constant). The number of failures in a given time will follow a Poisson distribution. Then the probability of more than N failures in time t is given by

$$P[n < N] = 1 - \sum_{n=0}^{N} \frac{(c\lambda t)^n}{n!} e^{-(c\lambda t)}$$

(8.23)

Since the number of failures is usually small, it is preferable to use this form rather than to attempt to sum the Poisson series to infinity.

> *Example: a system includes 4 printed circuit boards of a particular type. Each board has a MTTF of 12000 hours. Normally 2 boards are kept as spare and if a new board is required, the delivery time is 1 week. What is the probability that the system will fail and be without an immediately available spare during its 20 year lifetime?*

After each fault, there is a period of 1 week (168 hs) during which the system is vulnerable. If one fault occurs during this period, the remaining spare is used but if there is more than one fault, the system will break down.

In this case, c = 4; λ = 1/12 000   and t = 168

$$P[N > 1] = 1 - (1 + c\lambda t) e^{-c\lambda t}$$
$$= 1.51 \times 10^{-3}$$

In 20 years, the expected number of faults is

$$= 175\,000 \times 4 / 12\,000 \qquad = 58.3$$

and the probability of being without a spare

$$= 0.088$$

There is a 9% chance that the system will break down during its lifetime due to a fault on this type of PCB.

8.12.3   Reliability block diagrams

In more complex systems, the failure of a single component may not cause the system to fail. In this case, a diagrammatic form may be used to aid the reliability analysis. A reliability block diagram represents the reliability operation of a system by a set of blocks, each representing a component or subsystem. A signal is supposed to proceed through the blocks from input to output but the interconnections, instead of representing the operation of the system as in a normal block diagram, describe the reliability performance.

In the case of the simple system in which any component or subsystem failure causes the system to fail, the reliability block diagram (RBD) is represented by a simple series arrangement (Figure 8.10)

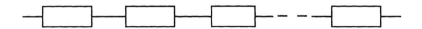

Figure 8.10   Reliability blocks in series

If there is a measure of redundancy in the system so that the system continues to operate when a particular component or subsystem has failed, a parallel connection is shown (Figure 8.11). Then if a component fails, there remains a path between input and output and this is interpreted as the system remaining in operation. In such systems it is assumed that if one subsystem fails, it does not have any further effect on the operation of the system. For example, human judgement may be used to ascertain which of two subsystems has failed. Alternatively a dual processor system may incorporate a self-checking mechanism such that if one processor should fail to give a predetermined output in response to a given stimulus, it is automatically removed from service.

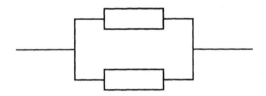

Figure 8.11 Reliability blocks in parallel

In other cases, the assumption may depend on the mode of failure of the component. For example consider two diodes connected in series (Figure 8.12). If the failure mode is such that a failed diode appears as a short circuit, then the assumption of parallel redundancy is justified. Even then there is at this stage the implied assumption that the reliability of the remaining diode is not reduced even though the peak inverse voltage is doubled.

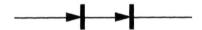

Figure 8.12 Diodes in series

For the system to fail in less than time t, both subsystems have to fail. Then for the system

$$F(t) = [1 - R_1(t)][1 - R_2(t)]$$

and the system reliability is

$$R_s(t) = 1 - F(t)$$

$$= 1 - [1 - R_1(t)] [1 - R_2(t)]$$

$$= R_1(t) + R_2(t) - R_1(t) R_2(t) \qquad (8.24)$$

If the subsystems have the same reliability,

$$R_s(t) = 2R(t) - R(t)^2$$

To find the system MTTF for a constant failure rate in the branches, we write

$$MTTF = \int_0^\infty R_s(t)\, dt$$

$$= \int_0^\infty \exp(-\lambda_1 t)\, dt + \int_0^\infty \exp(-\lambda_2 t)\, dt$$

$$\int_0^\infty \exp[-(\lambda_1 + \lambda_2)t]\, dt$$

$$= \frac{1}{\lambda_1} + \frac{1}{\lambda_2} + \frac{1}{\lambda_1 + \lambda_2}$$

$$(8.25)$$

Note that although we assumed a constant failure rate for the two branches, the reliability function for the system is not a simple exponential. The system MTTF is not then simply the reciprocal of the failure rate.

In general, for a n-branched system,

$$R_s = 1 - \prod_{j=0}^n (1 - R_j)$$

$$(8.26)$$

8.12.4   Majority decision redundancy

A particular application of parallel redundancy involves the use of an odd number (3 or more) subsystems carrying out the same function. A simple comparison circuit is used to give a system output which represents the majority decision of the subsystem outputs as shown in Figure 8.13. The operation of the total system is unaffected by any mode of failure of one subsystem. There is, of course, the assumption that the majority decision circuit itself does not fail.

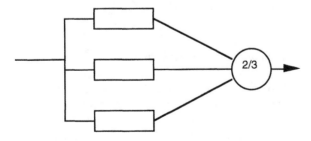

Figure 8.13   2-out of -3 majority decision redundancy

Then the system reliability

$$R_s(t) \quad = \text{P[2 units OK and 1 failed]} \quad + \quad \text{P[3 units OK]}$$

$$= 3R_1{}^2 (1 - R_1) + R_1{}^3$$

$$= 3R_1{}^2 - 2R_1{}^3$$

8.13   Markov modelling

As explained in Chapter 2, Markov models are very useful for giving a picture of changes in system state caused by random events in a continuum of time. They can therefore be applied in the study of reliability, particularly if the rate of random events or failures is constant. In simple systems such as those given above, a direct approach is possible but for more complex situations, Markov models provide a useful basis for analysis. The Markov model is also useful in cases where the probability of failure of a component or subsystem is not independent but depends on the states of other parts of the system. Two simple examples are given below of systems employing different types of redundancy.

8.13.1   *Parallel redundancy (hot standby)*

If we take the case of the two diodes in series (Figure 8.14), the short circuit failure of one diode may cause the remaining diode to have a higher probability of failure than hitherto because of the increased inverse voltage applied to it. The system may be modelled by a Markov diagram (Figure 8.14) in which nodes represent the states of the system and ties the failure rates. Under normal conditions, both components are operating and the probability of failure of one component in interval dt is $\lambda_1 dt$.

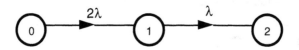

Figure 8.14   Markov representation of hot standby system

The transition probability between state '0' and state '1' is therefore 2 λ_1, since both components are in operation together. If one component should fail, the peak inverse voltage applied to the remaining diode is doubled and the failure rate is consequently increased to λ_2. The reliability of the system may then be calculated as follows:

The probability of transition from state '0' to state '1' in time **dt** is

$$dP_0 \ = \ - 2\lambda_1 P_0 dt$$

or

$$\frac{dP_0}{dt} \ = \ \dot{P}_0 \ = \ - 2\lambda_1 P_0$$

Similarly

$$\dot{P}_1 \ = \ 2\lambda_1 P_0 \ - \ \lambda_2 P_1$$

and

$$\dot{P}_2 \ = \ \lambda_2 P_1$$

These equations may be written in matrix form

$$\begin{bmatrix} \dot{P}_0 \\ \dot{P}_1 \\ \dot{P}_2 \end{bmatrix} = \begin{bmatrix} -2\lambda_1 & 0 & 0 \\ 2\lambda_1 & -\lambda_2 & 0 \\ 0 & \lambda_2 & 0 \end{bmatrix} \begin{bmatrix} P_0 \\ P_1 \\ P_2 \end{bmatrix}$$

that is

$$\dot{P}(t) \ = \ \mathbf{M} P(t) \tag{8.27}$$

The first-order differential equations may be solved to give the state probabilities as functions of time. Computer techniques are usually required for all but the simplest systems.

In the example given above, it may be shown that

$$P_0 \ = \ \exp\left(-2\lambda_1 t\right)$$

and

$$P_1 \ = \ \frac{2\lambda_1}{\lambda_2 - 2\lambda_1} \left[\exp\left(-2\lambda_1 t\right) - \exp\left(-\lambda_2 t\right) \right]$$

The reliability is then

$$R(t) = P_0 + P_1$$

$$= \frac{\lambda_2}{\lambda_2 - 2\lambda_1} \exp(-2\lambda_1 t) - \frac{2\lambda_1}{\lambda_2 - 2\lambda_1} \exp(-\lambda_2 t)$$

(8.28)

Note that if $\lambda_1 = \lambda_2$ this reduces to

$$R = 2 \exp(-\lambda t) - \exp(-2\lambda t)$$

which corresponds to Equation 8.24.

Examples of such systems include parallel generators or load sharing processors. Normally, all units are in operation but if one unit should fail, the full system load falls on the remaining units.

8.13.2   Cold standby operation

In this case, one unit is kept in operation and the other is not switched on until a failure is detected. This mode of operation is used mainly for mechanical systems such as generators to reduce wear. The tacit assumption is made that the probability of failure of the changeover mechanism is negligible. The Markov diagram for such a system is then shown in Figure 8.15.

The reliability may be found in the same way as before and is given by

$$R(t) = \exp(-\lambda_1 t) + \frac{\lambda_1}{\lambda_2 - \lambda_1} [\exp(-\lambda_1 t) - \exp(-\lambda_2 t)]$$

(8.29)

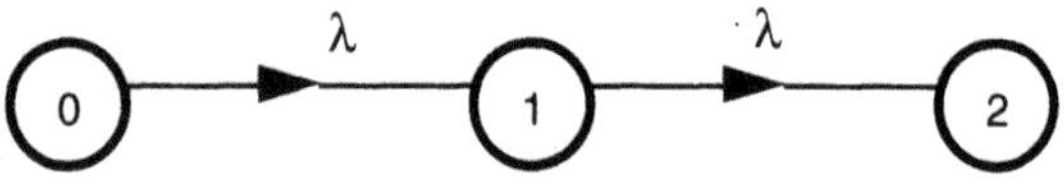

Figure 8.15   Markov representation of cold standby system

If $\lambda_1 = \lambda_2$, it is necessary to write the the exponential functions in the second term in series form. It may then be shown that

$$R(t) = (1 + \lambda) \exp(-\lambda t)$$

8.14   Fault-tree analysis

For large complex systems, it is often not immediately obvious how redundancy in different parts of the system affects the overall reliability. A useful diagrammatic aid to the reliability of such systems is known as a fault tree [4]. This is similar to a reliability block diagram but it provides a more compact indication of the effects of particular faults by the use of 'AND' and 'OR' gates. These are used purely for reliability assessment and have no connection with the logical operation of the system. Each individual 'event' (that is, a component or subsystem failure) is combined with other events in an OR gate if either event will cause failure at the next higher level, and an AND gate if all the events are required to produce higher level failure. This is continued up the 'tree' until the 'top event' corresponds to complete system failure.

Even a relatively simple case such as a processor system with cold standby becomes more complex when power supplies and changeover mechanism are considered. A typical fault tree for such a system is shown in Figure 8.16. The failure probability of the complete system is estimated by calculating the failure probability for each group of end events and so on up the tree. Computer programs are available for carrying out the calculations.

The advantage of fault tree analysis is that it enables the failure modes of extremely complex systems such as the processors used to control telephone exchanges or nuclear power stations to be examined logically in a manner that is familiar to engineers. However, it is basically a combinatorial system and it is difficult to provide adequate representation of sequential events such as caused by a dormant failure in a changeover mechanism. Another weakness of the technique is that common mode failures are not always revealed. For example, if several leads are run close together, a small fire can cause failures in all the leads. The result might be a major breakdown that would appear virtually impossible from the calculations.

8.15   Repairable systems

Most large systems can be repaired during their operating life normally by replacement of the faulty component or subsystem. This may be regarded as a 'birth and death' process although in the case of reliability studies, it would be more appropriate to speak of a 'death and rebirth' process. The analysis is carried out in the same way as for other birth and death processes such as queueing.

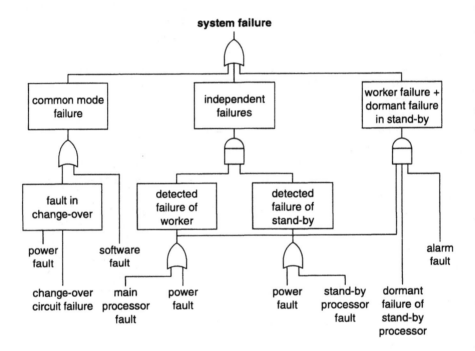

Figure 8.16 Fault tree representation of processor system with stand-by

Although we speak generally of repair, the complete process involves:

- detection that a fault has occurred;

- accurate diagnosis of the fault;

- getting maintenance personnel to the fault;

- obtaining the necessary spare units;

- replacing the faulty units;

- checking that the repair is satisfactory;

- restarting any units that have been closed down as a result of the fault;

- checking the overall operation of the system.

The time required for these operations is very variable and dependent on the maintenance policy adopted for the system. For example, some faults, if they are not catastrophic for the system, may remain undetected until the next maintenance check. In this case the probability function for the time to detect the fault would follow a rectangular distribution in the range 0 to T_m where T_m is the interval between maintenance checks. The total repair time might then be only slightly greater than the fault detection time. However, since reliability calculations are relatively crude we can assume that the repair time follows a negative exponential distribution so that it is consistent with a birth and death process. At least this simplifies the analysis.

Then

$$P \text{ [repair is completed in interval } \delta t] = \mu \delta t$$

The quantity μ is then known as the repair rate by analogy with the fault rate, λ. Both λ and μ are assumed to be constant. Furthermore, by analogy with the mean time to failure (MTTF) under constant failure rate conditions, the mean time to repair is simply the inverse of the repair rate.

$$MTTR = \frac{1}{\mu}$$

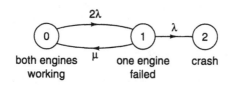

Figure 8.17 Markov representation of aircraft engine system

In some cases, repair is possible only under certain conditions. An example is a two-engined aircraft, which is normally capable of flying on one engine. Time is measured in flying hours and the MTTF should be many thousands of hours. The MTTR in this case would be taken as one half the mean flight time, since this is the average time to discover the fault. The actual time to repair the fault is not important in considering the system reliability. The Markov diagram for the system is given in Figure 8.17 and the transition matrix is

$$
\mathbf{M} = \begin{bmatrix} -2\lambda & \mu & 0 \\ 2\lambda & -\lambda-\mu & 0 \\ 0 & \lambda & 0 \end{bmatrix}
$$

(8.30)

The differential equations would then be solved in the usual way to give the probability of catastrophic failure P_2.

8.15.1 Availability

In the case of the aircraft, a failure of all the engines would result in a crash and the system would cease to be of future use. In many electronic systems, a complete system breakdown, although serious, would still allow recovery to a normal operating state. During the lifetime of the system, there may be several breakdowns followed by repair and we are interested in the fraction of the total time that the system is operational. This is known as the availability of the system. With modern systems, the availability may be very close to unity and it is sometimes more convenient to refer to the unavailability .

8.16 Redundant systems with repair

The availability of more complex systems such as those involving extended redundancy may be calculated by analysis of the Markov process as in teletraffic studies. We start by using the technique for the simple systems covered above.

8.16.1 Parallel redundancy with repair

The Markov diagram for a simple redundant system is shown in Figure 8.18. Under normal conditions, both subsystems are operating so that the fault rate is 2λ. When one subsystems fails, the fault rate for the remaining subsystem falls to λ. The repair transition probabilities depend on the maintenance policy. Here it is assumed that only one subsystem can be repaired at a time, so that the repair rate is μ irrespective of the number of subsystems that are faulty. In some situations, the provision of maintenance staff and the accessability of the equipment might permit two faults to be repaired simultaneously and the transition probability from state '2' to state '1' would be increased to 2μ.

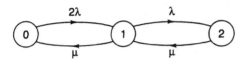

Figure 8.18 Markov representation of simple redundant system with repair

The 'probability flow equations' may be written

$$2\lambda P_0 - \mu P_1 = 0$$

$$2\lambda P_1 - \mu P_2 = 0$$

Then applying the normalising condition

$$P_0 + P_1 + P_2 = 1$$

and solving the equations as before we find the probability that the system will be unavailable

$$P_2 = \frac{2\lambda^2}{2\lambda^2 + \mu^2 + 2\lambda\mu}$$

$$(8.31)$$

8.16.2 Majority decision redundancy

A simple system may be analysed as in Section 8.12.4. The Markov diagram is given in Figure 8.19. This may be analysed in the same way as before and the availability of the system (equal to the sum of the state probabilities, P_0 and P_1) calculated. In such a simple case, it hardly seems worthwhile to invoke the Markov process when the analysis may be carried out by inspection. However, a practical majority decision system is rather more complex. Under normal conditions, if one subsystem should develop a fault, the system will continue to function and it will not be apparent that a fault has indeed occurred. It is therefore necessary to add a fault detection mechanism to indicate when the output of one subsystem is different from that of the others. A suitable arrangement is shown in Figure 8.20. A problem then arises if the fault detection circuit itself develops a fault. A simple manual check may be built in as shown and this is checked by a technician at regular intervals. The fault detection circuit is relatively simple and will normally have a much lower fault rate than that of the subsystems. However, the MTTR of the fault detector, which includes the fault detection time (that is, the time from the last manual check) may be much greater than the MTTR of the main units. The behaviour of the fault detection system may therefore have substantial impact on the overall system reliability.

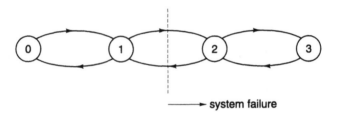

Figure 8.19   Markov representation of majority decision redundant system

The corresponding Markov diagram is shown in Figure 8.21. Some simplfying assumptions may be made depending on the maintenance procedures. In this case, it is assumed that in the case of both the alarm system and one of the main subsystems developing a fault, the alarm system is repaired first.

Typical values for the transition coefficients are:-

λ_1 =   fault rate of subsystem (10^{-4} h^{-1})

λ_2 =   fault rate of alarm system (10^{-5} h^{-1})

μ_1 =   repair rate of subsystem or alarm system if fault discovered immediately (0.25 h^{-1})

μ_2 =   repair rate of alarm system if fault discovery dependent on manual test (0.0125 h^{-1})

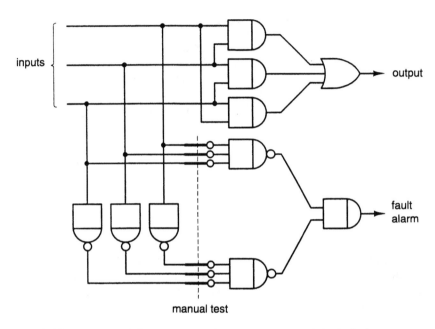

Figure 8.20   Two-out-of-three majority decision system with fault detection

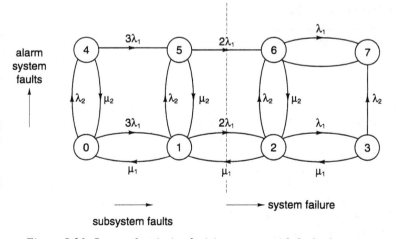

Figure 8.21   States of majority decision system with fault detection

The corresponding transition matrix is then

$$M = \begin{vmatrix} -3\lambda_1-\lambda & \mu_1 & 0 & 0 & \mu_2 & 0 & 0 & 0 \\ 3\lambda_1 & -2\lambda_1-\lambda_2-\mu_1 & \mu_1 & 0 & 0 & \mu_2 & 0 & 0 \\ 0 & 2\lambda_1 & -\lambda_1-\lambda_2-\mu_1 & \mu_1 & 0 & 0 & \mu_1 & 0 \\ 0 & 0 & \lambda_1 & -\lambda_2-\mu_2 & 0 & 0 & 0 & 0 \\ \lambda_2 & 0 & 0 & 0 & -3\lambda_1-\mu_2 & 0 & 0 & 0 \\ 0 & \lambda_2 & 0 & 0 & 3\lambda_1 & -2\lambda_1-\mu_2 & 0 & 0 \\ 0 & 0 & \lambda_2 & 0 & 0 & 2\lambda_1 & -\lambda_1-\mu_1 & \mu_1 \\ 0 & 0 & 0 & \lambda_2 & 0 & 0 & \lambda_1 & -\mu_1 \end{vmatrix}$$

$$(8.32)$$

The solution of the corresponding equations to obtain the state probabilities would then enable the unavailabilty to be found as the sum of the probabilities of states 2, 3, 6, and 7. Computer techniques would normally be necessary for such complex problems [5].

8.17   Software reliability

Strictly speaking, there are no software 'faults' in the sense that hardware faults are caused by component failures. Any malfunction of the software is essentially due to a design error. Large software systems almost inevitably contain a substantial number of incipient software faults or 'bugs'. Some of these may not become evident until the system has been operated for some time. The reason they come to light is that a particular sequence of inputs has not occurred previously or that a minor malfunction has not been noticed by the user. Failure or hazard, whether caused by hardware or software is essentially the same. It is the inability of a unit to continue to perform its specified function.

It is virtually impossible to test fully some large systems since the number and timings of all possible input sequences is far too great. All that can be done is to run the system for some time under controlled test conditions and monitor the operation. This means that systems put into use often contain software bugs but, if the testing has been carried out properly, the system failures should not occur too frequently.

As for hardware, the reliability of the system can be improved by careful design [6]. This involves:

- requirements: formal methods such as the CCITT 'specification and description language' (SDL) impose a discipline on the requirements;

- design: dividing the system into well defined modules facilitates coordination within the design team;

- coding: a good compiler and possibly a static analyser will examine the program for a number of features including semantics, syntax, data low etc;

- qualtity assurance: independent tests designed to expose any weaknesses in the system;

- fault recovery: even without redundancy, software can be designed so that a fault indication is printed out and the system continues to operate.

8.17.1 Software testing

Before being released for use, most software systems undergo a period of testing. A separate processor is often used to simulate the inputs to the software and to observe the reactions. At the beginning of the test period, faults will occur relatively frequently. After each fault is revealed, the time is noted and the clock is stopped until the bug is found and corrected. The testing is then restarted until the next bug appears. If the software has been well designed, the intervals between successive stoppages should increase. (It is not unknown in testing badly designed software for the fixing of one bug to result in other bugs being introduced!)

From the data produced during testing, an estimate may be made of the parameters of an assumed distribution. A rough prediction of the performance when the software is released to the user may then be made. If it is assumed that further software fixes need not be considered during the relevant operational period, then it is only necessary to find the failure rate at the end of the testing period. The reliability during operation is then found from Equation 8.8, assuming a constant failure rate as shown in Figure 8.22. This assumption is usually made when we are concerned with, say, a short period of acceptance trials and a management decision on whether or not to release the software will depend on the probability of passing the acceptance test. An alternative scenario is that the fault rate continues to fall during the operational period. This of course could only occur if there are several installations of the same software. It is then assumed that bugs being cleared on other installations will result in modifications being made in the installation being observed with no inconvenience to the user. This will be reflected in the failure rate continuing to fall in accordance with the Weibull equation during the operational phase as shown by the broken line of Figure 8.22.

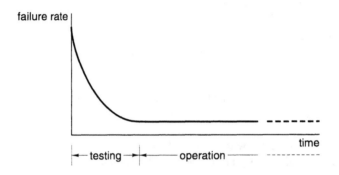

Figure 8.22   Failure rate during software testing

> *Example: after 1000 hours run-time testing, the bug rate is found to follow a Weibull distribution with constants* $m = 0.35$ *and* $\theta = 7500$. *What is the probability that no further bugs will occur during the first three months of operation?*

Assumption 1 - constant failure rate during operational phase

Then failure rate

$$\lambda\,(1000) \;=\; \frac{0.35}{7500}\left[\frac{1000}{7500}\right]^{-0.65} \;=\; 1.73 \times 10^{-4}$$

At this constant rate, the reliability at 3 months (2190 hs)

$$R(2190) \;=\; \exp(-0.379) \;=\; 0.685$$

There is an 68.5% chance that no further bugs will occur during the first three months.

OR

Assumption 2 - failure rate continues to fall during operational phase

Then

$$R(1000) = \exp\left[-\left(\frac{1000}{7500}\right)^{-0.35}\right] = 0.610$$

and

$$R(3190) = \exp\left[-\left(\frac{3190}{7500}\right)^{-0.35}\right] = 0.476$$

Hence reliability at 2190 h in operational phase

$$= \frac{R(3190)}{R(1000)} = 0.781$$

the reliability during the first three months of operation is now 78%.

The choice between Assumption 1 and Assumption 2 will depend on the conditions under which the software is put into operation.

8.18 References

1   STITCH, M. *et al* : 'Microcircuit accelerated testing using high-temperature operating tests', *IEEE Trans. on Reliability,* **R-24**, (1975) pp. 238–250

2   'Reliability prediction for electronic systems', *US MIL-HBK-217,* National Technical Information /Service, Springfield Va. (1986 with later updates)

3   'Reliability in systems, equipments and components', *BS 5760,* British Standards Institution

4   'Fault tree handbook' *NUREG-0492,* US Nuclear Regulaitory Commission (1981)

5   BAJPAL, A .C. *et al* : 'Engineering mathematics', Wiley (1974)

6   'Software quality assurance program requirements' , *US MIL- STD-52779A,* National Technical Information /Service, Springfield Va. (1987)

Chapter 9

Miscellaneous examples of performance engineering

9.1 Introduction

Although performance engineering originated in the study of telephone calls randomly originated by subscribers, the applications have been greatly extended in recent years. There are many applications outside the field of information engineering such as the study of road traffic, particularly the problems of queuing at roundabouts and obstructions. These are best dealt with in books and papers pertaining to the applications. Even within the field of information engineering, it would be impossible to cover the complete range of applications of performance engineering in a book of this size. Chapter 8 has covered the important field of reliability and this chapter gives a few more examples to illustrate the techniques.

9.2 Tolerances

In mechanical engineering, tolerances are usually regarded as absolute limits and component parts are made to fall within the tolerance limits. The statistical aspects are not usually considered except in production engineering where, for example, sampling techniques may be used to decide when a production line may be about to produce an excessive proportion of components outside the set tolerance. In electronic engineering, on the other hand, most components are produced with relatively wide tolerances and quite often the equipment design is sufficiently robust to operate satisfactorily even with components on the limits of their tolerances. In the past, for cases where the unit performance was critical, the problem was often overcome by fitting one or more pre-set adjustments that were set so that the unit operated at the ideal design point. In a large system it was then hoped that temperature variations and ageing did not cause so much drift in the component parameters that the system failed to operate satisfactorily.

In modern electronic systems design, the pre-set adjustment is frowned upon. Components can be manufactured to closer tolerances and more is understood about the effects of temperature. It is then important that units manufactured on a production line should operate within the prescribed limits and that the probability of a unit falling outside the limits should be acceptably small. However, economic design demands that the tolerances placed on the components should be as wide as possible. Probability techniques can then be used to assess the probability of a unit failing to perform within its specification and a decision can be made as to whether the design is satisfactory. For example, if only 10 000 units are to be manufactured, it would probably be acceptable if the probability of a reject due to component tolerances were 10^{-5}, bearing in mind some rejects may be caused by faults in assembly and the like.

9.2.1 Component values

In the manufacture of electronic components, small variations in the production line result in variations in the value of the components being produced. Since the process variations are usually random in nature, it is reasonable to suppose that the values are normally distributed about a mean that is the nominal value of the component. Unfortunately, in some cases a manufacturer selects the closer toleranced components to sell at a higher price as shown in Figure 9.1.

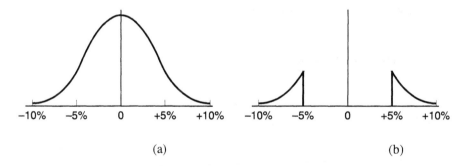

(a)   (b)

Figure 9.1 Distributions of component values

    (a) output of production line
    (b) after selection of close-toleranced components

Assuming that the tolerances of the components are known, how do they affect the performance of the unit or system? The performance of a unit often depends on the product or quotient of component values, so that we have to consider the distribution of the product or quotient of a number of normally distributed functions. It can be shown (although the proof is surprisingly complex) that the spreads of the component distributions are small and there is no correlation between them, the resultant distribution is also normally distributed with standard deviation given by

$$\sigma_T = \sqrt{\sum_{r=1}^{n} \sigma_r^2}$$

(9.1)

As an example, consider the transistor amplifier shown in Figure 9.2.

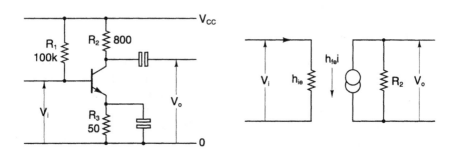

Figure 9.2   Simple transistor amplifier

From the simplified equivalent circuit

$$gain = \frac{V_0}{V_i} = \frac{h_{fe}}{h_{ie}} R_2$$

If $h_{fe} = 250$, $h_{ie} = 1000$ ohms and $R_2 = 800$ ohms, then

$$gain = 200 \text{ (or 46 dB)}$$

Suppose the transistor parameters are normally distributed with standard deviations of 37.5 (15%) and 120 (12 %) respectively and that the resistor R_2 also has a normally distributed value with standard deviation 10%.

The standard deviation of the gain distribution is then

$$\sigma_T = \sqrt{(15^2 + 12^2 + 10^2)} = 21.7\%$$

The probability that the gain will be less than 43 dB (141) then corresponds to $1.36\sigma_T$. From tables of the normal distribution, this equals 8.7%. If such amplifiers were produced in quantity and were required to have a gain not less than

43 dB, 8.7% of them would be rejected at the unit test stage – a clearly unsatisfactory situation.

If circuits are constructed on a single chip, there is a high degree of correlation in the values. If in the above example, the variations in the parameters h_{ie} and R_2 have a coefficient of correlation, ρ, then the standard deviation of the gain distribution becomes

$$\sigma_T = \sqrt{15^2 + 12^2 + 10^2 - 2\rho(12 \times 10)}$$

If the coefficient of correlation is say, 0.8 then the standard deviation of the gain distribution is reduced to 16.5% and the probability of an amplifier having a gain less than 43 dB is reduced to 3.7%

9.2.2   Timing tolerances

In many digital systems, pulses pass through several logic gates in series. The switching delays in the gates are subject to tolerances with the result that after passing through several stages, the pulses become distorted because of the multiple fan-out and fan-in at each stage. The usual practice is to use a central clock and to retime the pulses at suitable intervals. If after passing through a particular combination of logic gates, the pulses are so distorted that the retiming instant falls outside the signal pulse, errors will occur. The problem is to ensure that when the system operates at the desired clock rate the probability of error is negligible.

A typical example is shown in Figure 9.3 for a system working at a 20 MHz clock rate and 50% duty cycle. The timing is shown in Figure 9.4 and the problem is to find the tolerances on the logic gate switching times that will permit satisfactory operation. Fixed delays can be inserted where necessary.

Two approaches are possible:

(i)   Worst case design
The tolerances are assumed to be as follows:

clock distribution	± 2 ns
logic gate turn-on time	$\pm t_1$
logic gate turn-off time	$\pm t_2$
retiming toggles switching time	$\pm t_2$

clock
distribution

Figure 9.3   Path through a clocked digital system

A mis-selection will occur unless the clock pulse for the second retiming toggle always falls within the pulse after distortion by the logic gates. By inspection of the diagram, it will be seen that the condition for correct selection is

$$3t_1 + 5t_2 + 8 \leq 25$$

Note that any fixed switching times can be compensated by the delay network in the pulse distribution. We are only concerned with the tolerances. If, for example, $t_2 = 2\,t_1$, then the tolerance on switch-on time must be within ± 1.3 ns and on the switch-off time within ± 2.6 ns.

(ii) Probabilistic design
In this case we assume that the switch-on time at each of the logic gates is normally distributed with standard deviation σ_1. The switch-off time for the gates and the switching time for the toggle is similarly distributed with standard deviation σ_2. Worst case conditions are assumed for the clock pulse distribution. If we wish the probabality of retiming failure to be very small, say 10^{-7}, from tables of the complementary normal distribution function, this corresponds to 5.2 standard deviations. Then to achieve the required probability of timing failure,

$$5.2 \left(\sqrt{3}\, \sigma_1 + \sqrt{5}\, \sigma_2\right) + 8 \leq 25$$

Here again, if $\sigma_2 = 2\sigma_1$ then the maximum value of σ_1 will be 0.52 ns and that of σ_2 will be 1.04 ns.

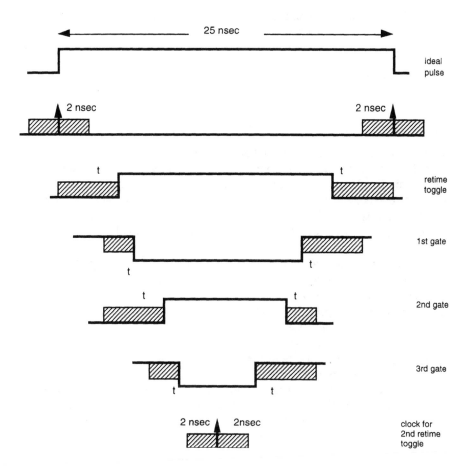

Figure 9.4   Timing tolerances through clocked digital system

9.3   Transmission systems

Although performance engineering is normally associated with switched networks, there are many cases in which the techniques may be applied to transmission systems. One of the earliest applications was in connection with long-distance (analogue) transmission. The 4-wire part of the system included several 'repeaters' (amplifiers) and it was essential to ensure that the total gain round the loop consisting of the amplifiers and the two 2-wire to 4-wire hybrids did not exceed unity, otherwise the system would break into oscillation. The gains of the repeaters were set so that this should not occur. However, tolerances in the adjustment of repeater gain and drift with time meant that the system had to be designed to allow an adequate saftety margin against oscillation. The usual procedure was to assume a normal distribution of repeater gains and to specify the mean gain so that the probability of oscillation was extremely small.

The design of digital transmission systems also offers some scope for the use of a probabilistic approach.

9.3.1   Repeated transmissions

When digital signals are sent over fading transmission channels involving, for example, HF or mobile radio links, error rates can be very high. One way of overcoming this is to repeat each message one or more times. If a suitable error detecting code is used, it is possible to ascertain the correct version of the message. An alternative is to send the message an odd number of times and use majority decision logic on each bit to decide which version is most likely to be the correct one.

Consider for example, a message which is sent twice over a fading channel. The fades may persist for several bits, so that if the message is repeated immediately after the first transmission, both versions of the message may be effected. This may be overcome by arranging for other messages to be sent between the first and second transmissions, a process known as interleaving.

A convenient model for the radio channel is a two-state discrete-time Markov 'chain' , operated at the bit-rate of the signal (Figure 9.5). In state '0' the signal strength is supposed to be above the noise or interference threshold and bit errors are very unlikely to occur. In state '1' on the other hand, the bit error rate becomes very high. For simplicity, let us assume that if during a transmission the model is in state '0' , the message will be received correctly and if for any part of the message, the model is in state '1', the message will be in error.

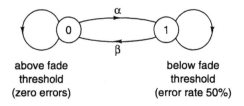

Figure 9.5   Markov model of fading on mobile radio link

The state probabiliity, p_1 represents the overall bit error ratio and p_0 is simply $1-p_1$. The probability 'flow' is then the inverse of the number of times the signal fades during a bit interval, that is

$$P_0 \, \alpha \; = \; P_1 \, \beta \; = \; \frac{h}{b}$$

(9.2)

where h = number of fades per second
b = bit rate of signal.

In the first case, we will assume that the two transmissions are interleaved so that a fade that affects one transmission will not affect the other. For a single isolated transmission of N bits to be received correctly, the model must be in state '0' for the first bit and remain in that state for the remaining N–1 bits. The probability of this happening is then

$$P_1 \; = \; P_0 (1-\alpha)^{N-1}$$

(9.3)

Then since the two transmissions are isolated, the probability of the message being received correctly is

$$
\begin{aligned}
P_{cv} \; &= \; 1 - [1 - P_1]^2 \\
&= \; 1 - [1 - P_0(1-\alpha)^{N-1}]^2
\end{aligned}
$$

(9.4)

For the second case, it is assumed that the second transmission follows immediately after the first. The effects of autocorrelation in the fading (see chapter 2) means that what happens to the second transmission depends on the condition of the model at the end of the first transmission. There are four possibilities:

(a) both transmissions received correctly;
(b) first tramsmission correct and second in error;
(c) first transmission in error and second correct;
(d) both transmissions in error.

If the first transmission is received correctly, it does not matter whether the second transmission is correct or not. Then the combined probability of (a) or (b) is simply the probability of a single isolated transmission being received correctly, P_1 as given by Equation 9.3.

In the consideration of (c), the behaviour of the second transmission will depend on the condition in which the last bit of the first transmission was received. The probability that the last bit of the first transmission is in condition '0' (correct) is p_0 but this is irrespective of whether the first transmission was correct or in error. The combined probability that the last bit of the first transmission is in condition '0' and the first transmission in error is then $p_0 - P_1$. Under these conditions, for the second transmission to be received correctly the model must then remain in condition '0' for N bits. The probability of this is

$$P_2 = (p_0 - P_1)(1 - \alpha)^N$$

(9.5)

The other possibility in connection with (c) is that the last bit of the first transmission is received in condition '1'. The probability of this is simply p_1 since the first transmission must have been in error. Under these conditions, for the second transmission to be correct, the model must transit to condition '0' for the first bit and remain in that condition for the remaining N – 1 bits. The probability of this is

$$P_3 = p_1 \beta (1 - \alpha)^{N-1}$$

(9.6)

The total probability of the message being received correctly is then found from Equations 9.3, 9.5 and 9.6

$$P_{cu} = P_1 + P_2 + P_3$$

$$= (1 - \alpha)^{N-1} \left\{ p_0 [2 \, \alpha - (1 - \alpha)^N] + p_1 \beta \right\}$$

(9.7)

> *Example: a digital channel operates at 600 bit/s and the overall bit error ratio is 5%. The signal fades below the error threshold on average 2 times per second. What is the probability that a 20 bit message will fail if*
>
> *(a) the message is sent once only;*
> *(b) the message is sent twice with a gap of several seconds between the first and second transmissions*

(c) the message is sent twice with the second
transmission following immediately after the
first.

The fading model in this case (see Figure 9.5) has constants

$$p_1 = 0.05$$
$$p_0 = 0.95$$
$$\alpha = 0.00351$$
$$\beta = 0.06667$$

(a)

The probability that an isolated transmission will be received correctly

$$= \; 0.95 \, (1 - 0.00351)^{19} \; = \; 0.8886$$

Probability of incorrect message = 0.111

(b)

Probability that both transmissions will be in error $= 0.111^2$

Probability of incorrect message = 0.012

(c)

Probability that first transmission in error and 20th digit received when
model is in condition '0' $= 0.95 - 0.8036 = 0.1464$

Conditional probability then that second transmission correct
$$= \; 0.1464 \, (1 - 0.00351)^{20} \; = \; 0.0572$$

Conditional probability that 20th digit received when model in condition
'1' and second transmission correct
$$= \; 0.05 \times 0.06667 \, (1 - 0.00351)^{19} =$$
$$0.0031$$

Hence total probability that message correct $= 0.8886 + 0.0572 +$
$$0.0031 = \; 0.9489$$

Probability of incorrect message $= 0.051$

This example emphasises the advantages of interleaving. The 11% message error
rate is only reduced to 5% if the repetition takes place immediately after the first
transmission. By introducing interleaving, the error rate is reduced to just over 1%.
However, the full results of interleaving are realised only if the second transmission

is completely independent of the first. In other words, the second transmission should not take place until the autocovariance of the fading model has fallen virtually to zero.

9.3.2 Performance of error detecting and correcting codes

In this example we consider the use of a slightly more complex model of the transmission channel in estimating the performance of error detecting and correcting codes. Errors may be caused by thermal noise which very occasionally exceeds the amplitude of the signal by impulse noise or by the fading of a radio signal. In the first two cases, the duration of the error is very short, usually only one or two digits. Fading on the other hand often causes errors to occur in bursts, so that several successive digits are in error.

A simple model of a digital channel subject to errors is shown in Figure 9.6. The model is usually in state '0', corresponding to zero errors but occasionally it moves to state '1' (sporadic error) or state '2' (error burst). A further state representing a combination of sporadic and burst errors might be postulated but if the probability of errors is very small, this state can be neglected. It is assumed that since binary signals are being transmitted, there is still a 50:50 chance of a correct digit being received even when the model is in the 'error' conditions of states '1' and '2'.

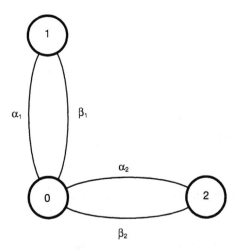

Figure 9.6 Markov states for single errors and error bursts

A wide variety of error detecting and correcting codes are available and are described in textbooks. The performance of such codes may be estimated using a model for the transmission link such as that described above. The calculations are usually straightforward but the main problem is in deciding which approximations are appropriate. In the following example, values are given to the transition coefficients so that the validity of the approximations may be checked.

The simplest form of error detection is a parity check, for example a single parity bit might be added to each 7-bit word. Consider first the effect of sporadic errors involving a transition from state '0' to state '1'. A shift from state '0' to state '1' may occur with probability α_1 at any of the first seven bits and the double error is possible if it remains in state '1' for a further bit interval, with probability $(1 - \beta_1)$. We may neglect what happens after that. As explained above, even if the model is in the error condition, the probability of the bit being in error is 0.5. Hence the probability of an undetected error is

$$P_{ut} = 7 \times 0.5^2 \, \alpha_1 (1 - \beta_1)$$

(9.8)

Substituting values from the model,

$$P_{ut} = 7 \times 10^{-7}.$$

For errors caused by interference involving a transition from state '0' to state '2', the situation is more complex since there is a significant probability that the model will remain in the error condition for several bit intervals. Suppose the model changes to state '2' during the first bit of a word. We then have to consider the probability of the error conditon persisting for a further 1,2 ......7 bits. During that time, an undetected error will occur if 2 of the bits are in error and the rest correct. Neglecting the possibility of 4 or more bits in error and considering each possiblity in turn, the probability of undetected error is

$$\alpha_2 \beta_2 \left[0.5^2 \left(\frac{2}{2} \right) (1 - \beta_2) + 0.5^3 \left(\frac{3}{2} \right) (1 - \beta^2 + \dots + 0.5^8 \left(\frac{8}{2} \right) (1 - \beta_2)^7 \right]$$

(7 terms)

A similar series results if the model changes to state '2' during the second bit of the word except that the series consists of only 6 terms and so on. The total probability of undetected error is

$$P_{ub} = \alpha_2 \beta_2 \sum_{k=1}^{7} (8-k) \times 0.5^{k+1} \left(\frac{k+1}{2} \right) (1 - \beta_2)^k$$

(9.9)

Substituting values from the model,

$$P_{ub} = 1.2 \times 10^{-5}.$$

In this case, although the burst errors occur much less frequently than isolated errors due to thermal or impulse noise they are more likely to cause undetected errors.

9.4   Mobile radio example

An example of the interaction between random user behaviour and natural phenomena occurs in cellular mobile radio systems. The basis of cellular systems is the use of relatively small service areas or cells served by low-power transmitters. The channel frequencies can then be re-used at distances such that the probability of interference is acceptably low. In this way, the maximum number of simultaneous calls may be made for a given amount of frequency spectrum. Because of the small cells, as a mobile unit moves from one 'cell' to another, the call must be handed over to a free channel in the new cell. This process takes place automatically and the break in transmission should be so short that it is not perceptible to the user.

A problem arises if all the available channels in the new cell are already occupied. In this case, it might appear that the call would have to be dropped as the mobile unit moved over the cell boundary. This would be particularly annoying to the user, far more so than having a new call blocked. Fortunately, the situation is made easier by the fact that radio contact can be maintained for some distance after the mobile has crossed the cell boundary. Obviously there is a limit to this and the object of the exercise is to find the probability of a channel being cleared in the new cell before the signal from the old base station becomes unusable.

Radio propagation is a complicated subject but for the purpose of this example, it may be assumed that the mean received signal power varies as $1/d^{\alpha}$, where d is the distance from the tranmitter and the constant, α has a theoretical value of **4** but in practice is about **3.6**. The signal undergoes fading due to buildings and other obstructions and it is found that the distribution of received signal power about the mean follows a log-normal distribution, that is the distribution is normal or Gaussian when the signal levels are measured in decibels. The signal also undergoes fading due to interactions between multiple ray paths but this fading is normally very rapid and need not concern us here. A fuller account of mobile radio propagation is given in books by Parsons and Gardiner and by Macario (ed.) [1, 2].

In this simplified example, we consider a mobile moving from one cell to another along a line joining the cell centres (Figure 9.7). We wish to find the probability that, during the time that all channels in the new cell are occupied, a call has to be dropped because it is not possible to achieve the handover. It is assumed that a call waiting for handover has priority over any new calls that arise.

The parameters of the system are as follows:

Physical
radius of cells	=	10 km
vehicle speed (constant)	=	14 m s^{-1}

Traffic
no. of channels/cell	=	10
Loading	=	5.1 Erlangs
Call blocking probability	=	0.0203
Mean call holding time	=	150 s

Radio

received signal strength   = $-36 \log d$ dB (relative to arbitrary level) (d = dist. from transmitter in m)

level below which signal is unusable (error threshold)   = -160 dB

standard deviation of log-normal fading   =   7 dB

mean distance between deep fades   =   500 metres

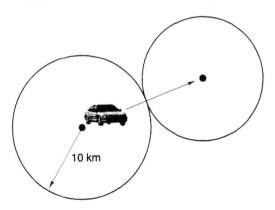

Figure 9.7   Mobile crossing cell boundary

To tackle the problem, we divide the path into 200 m sections, starting 100 m inside the old cell. At the mean point of the first section (i.e. on the boundary of the cell), the mean signal strength is -144 dB relative to the arbitrary level. This is $160 - 144 = 16$ dB or 2.286 standard deviations of the log-normal fading distribution above the error threshold. From the tables of the normal distribution, the probability of the signal falling below the error threshold is 0.011. The fading model used is similar to that discussed in Section 9.3.1 (Figure 9.5). State '0' represents the condition when the signal is above the error threshold (probability $p_0 = 0.989$) and state '1' the state when it falls below (probability $p_1 = 0.011$). The model is triggered at 1 m intervals. The transition coefficients, α and β may be found from

$$\alpha \quad = \quad \frac{1}{500 \; \pi_0}$$

$$= \quad 0.00202$$

and

$$\beta \quad = \quad \frac{1}{500 \; \pi_1}$$

$$= \quad 0.182$$

since the mean distance between deep fades is 500 m.

From the model, the probability that a fade will occur in 1 m is simply α. If a fade does occur, the probability that it will persist for 2 s or more is the probability that the model will remain in the '1' state for 27 or more metre intervals followed by a return to the '0' state, that is

$$\beta[(1-\beta)^{27} \quad + \quad (1-\beta)^{28} \quad + (1-\beta)^{29} \quad + \dots \dots \; + \quad]$$
$$= \quad (1-\beta)^{27}$$

Then the probability that the signal will enter a deep fade is

$$p_f \quad = \quad \alpha(1-\beta)^{27}$$

$$(9.10)$$

However, while the mobile is travelling along the 200 m section, it is possible that a call will be cleared in the new cell and a handover can take place. The probability of a call being cleared in the new cell during the time the vehicle travels 1 m is

$$p_c \quad = \quad \frac{10}{150 \times 14} \quad = \quad 4.76 \times 10^{-3}$$

The call will be cleared during the metre interval if the signal enters a deep fade and if the call has not already been cleared. Then the probability of a call being dropped during the 200 m section is

$$
\begin{aligned}
P_d &= \sum_{k=0}^{199} P_f \left[(1 - P_c)(1 - P_f) \right]^k \\
&= P_f \frac{1 - \left[(1 - P_c)(1 - P_f) \right]^{200}}{1 - (1 - P_c)(1 - P_f)}
\end{aligned}
$$

$$(9.11)$$

A similar fading model may be constructed for the next 200 m section of the route. The values will of course be different since the mean signal strength is decreasing and therefore the probability of fading below the error threshold will increase. Furthermore, it is possible that the call will have been already handed over or dropped due to fading in the first sections. The probability of this happening is

$$
P_s = (1 - P_d)(1 - P_c)^{200}
$$

$$(9.12)$$

The probability that the call will be dropped in the second 200 m section will then be $P_s P_d$ and in the third section, $P_s P_d^2$ and so on.

The probabilities that the call will be dropped due to a deep fade occuring before a free channel becomes available in the new cell are plotted in Figure 9.8. It will be seen that the chances of a dropped call fall rapidly as the mobile moves over the cell boundary. The total probability of a dropped call from the data given is 3.4×10^{-3}. Since, with the given loading, the blocking probability is 0.02, this means that the overall probability of dropping a call on handover is 7×10^{-5}, which may (or may not!) be considered satisfactory.

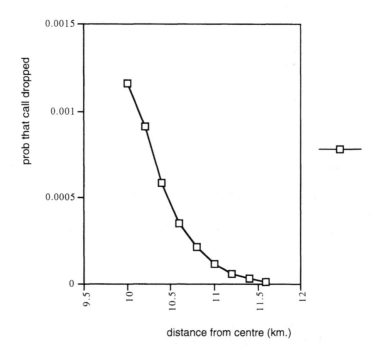

Figure 9.8   Probability of dropped call due to deep fade

The treatment given above is, of course, greatly simplified. A full assessment of the performance would take into account a range of traffic speeds and directions. Nevertheless, a few calculations would enable the range of call dropout probabilities to be calculated for typical situations and a decision made as to whether or not the system design was satisfactory. They would also enable the designer to test for sensitivity of the system performance to changes in the assumed parameters. For example, if the calculations on the above design are carried out under the assumption that the standard deviation of the log-normal fading were 8 dB instead of 7 dB, the performance is found to be very much worse.

9.5 References

1 PARSONS, J. D. and GARDINER, J. G.: 'Mobile communication systems', Blackie/Halstead Press (1989)

2 MACARIO, R. C. V.: 'Personal and mobile radio systems', Peter Peregrinus, on behalf of the IEE (1991)

3 GILBERT, E. N.: 'Capacity of a burst noise channel', *Bell System Tech Journal*, **39,** (1960) pp.1253–1265

Index

Accept-reject method, 68
Acknowledgement, 219-221, 226, 228, 230, 233, 234
  negative, 221
Additive congruential generator, 55
Arrival pattern, 94, 95
Arrival process, 78, 95, 96, 98, 99, 103-105, 117, 121, 126, 132, 153, 203, 207
  deterministic, 95
  general independent, 96
  Poisson, 38, 39, 41, 42, 44, 46, 67, 68, 95, 96, 102, 105, 106, 108, 117, 121, 132, 189, 203, 240, 244, 245, 250, 255, 263, 270, 282
Arrival rate, 67, 68, 81, 89, 95, 96, 100, 104, 110, 112, 127, 129, 131, 132, 155, 157, 160, 164, 170, 176, 191, 201, 208, 209, 211, 212, 247, 249
Assertions, 73, 79, 88
Asynchronous transfer mode (ATM), 97, 121, 201, 235, 236, 250, 251, 253, 256-258
Audio, 235
Autocorrelation, 43, 44, 242, 251, 255, 309
Autocovariance, 43, 44, 243-246, 312
Automatic repeat request (ARQ) protocol, 218
Autoregressive process, 243
Availability, 6, 15, 162, 163, 262, 264, 294, 295

Balance equations, 48, 132
Baskett, Chandy, Muntz and Palacios (BCMP) theorem, 134
Bathtub curve, 270, 272
BCMP
  network, 134, 140, 145, 150
  theorem, 134, 135
Bernoulli distribution, 19, 21, 28
Binomial distribution, 20, 28, 30, 107, 166, 184, 191, 237, 241, 245, 247, 280
Birth and death process, 44-47, 108, 116, 135, 291, 293
Bit-error ratio (BER), 4, 15, 217, 222, 308, 310
Bit-rate profile, 242
Blocking, 78-80, 83-86, 102, 109, 152, 162-165, 171-173, 179, 180, 182-189, 191, 195, 196, 198, 200, 214, 215, 236, 314, 315, 317
  downstream, 215
  upstream, 214, 215
Bottleneck, 137-139
Broad-band integrated services digital networks (BISDN), 235
Broadband network, 201
Bug, 268, 298-300
Burn in, 260, 268, 278, 281, 282

Call
  dropped, 317, 318
Carried load, 101, 105
Cell, 4, 201, 235-238, 250-253, 255, 256, 258, 314-317

Cell transmission interval (CTI), 237, 238, 251, 252, 255
Censored data, 275
Central server, 147-150
Chi-squared test, 60-62
Circuit switching, 200, 201, 236
Closed network, 128, 136, 138, 144, 147
  separable, 140
Component
  electronic, 6, 279, 281, 282, 303
  reliability, 274
  tolerance, 303
  value, 63, 303
Composition method, 63, 66
Compound randomness, 28
Conditional distribution, 37, 110
Conditional probability, 11, 46, 105, 107, 113, 186, 281
Conditional waiting probability density function, 204
Confidence intervals, 50, 75-77, 280, 281
Congestion, 96, 99-102, 225, 226, 234
Connectionless, 200, 201
Connection-oriented, 200, 201
Constant failure rate, 46, 268, 272, 274-276, 283, 286, 287, 293, 299, 300
Contention, 126, 136, 147, 149-151, 235, 236
Convolution, 26, 27, 141-145
  algorithm, 141-145
Correlation, 42, 43, 50, 60, 201, 303, 305
  cross, 42
Cost effectiveness, 259, 261
Covariance, 42, 43, 62, 242

Data
  censored, 275

Data circuit-terminating equipment (DCE), 225, 257
Data link, 110, 201, 218
Data sources, 240
Datagram, 200
Data-terminal equipment (DTE), 225, 257
Debugging, 50, 73, 75
Degrees of freedom, 61, 62, 76
Delay
  link, 202, 213, 222
  message, 202, 208, 211-213
  probability, 112, 113, 133, 152, 167, 172, 237
  propagation, 219, 220, 222, 234, 248
  round-trip, 219, 222
Derating, 260
Design, 1-6, 50, 90, 152, 164, 240, 256, 259, 260, 261, 264, 268, 274, 282, 283, 298, 299, 302, 305, 306, 308, 318
  probabilistic, 306
  tools, 6
  worst case, 305
Distribution
  Bernoulli, 19, 21, 28
  binomial, 20, 28, 30, 107, 166, 184, 191, 237, 241, 245, 247, 280
  conditional, 37, 110
  equilibrium, 47, 49
  Erlang, 113
  exponential, 35, 37, 38, 41, 46, 63, 67, 68, 99, 105, 113, 114, 132, 134, 145, 157, 169, 190, 201, 203, 207, 208, 226, 277, 293
  function, 31-33, 60, 61, 63, 64, 237, 264, 270, 273, 277, 306
  gamma, 35
  Gaussian, 34
  geometric, 23, 110, 191

joint probability, 25, 132, 140
lognormal, 314
negative-exponential, 35, 37, 38, 41,
    46, 51, 63, 64, 67, 68, 96, 99,
    105, 106, 113, 114, 132, 134,
    145, 155, 157, 161, 169, 170,
    189, 190, 201, 203, 206-208, 226,
    237, 277, 293
normal, 34, 35, 65, 75-77, 241, 244,
    270, 271, 280, 303, 304, 306,
    308, 314, 316
Poisson, 22, 39, 41, 51, 66-68, 105,
    118, 120, 121, 157, 161, 170,
    217, 226, 237, 284
probability, 16, 19, 34, 51, 63, 66,
    73, 101, 117, 144
steady-state, 47
Student-t, 75, 76
uniform, 51, 62, 63, 64, 155
waiting-time, 102, 103, 113, 114,
    120, 133, 206
Weibull, 271-273, 276-278, 300
Downstream blocking, 215
Dropped call, 317, 318

Electronic component, 6, 279, 281,
    282, 303
Equilibrium distribution, 47, 49
Erlang, 2, 5, 6, 78, 89, 95, 109, 111,
    113, 120, 133, 161, 162, 164, 184,
    192, 193
Erlang distribution, 113
Erlang's delay formula, 133
Error detecting codes, 308, 312, 313
Event queue, 70, 79, 81, 86-88
Event-based simulation, 70, 72, 77
Expectation, 16, 17, 22, 24-26, 32, 33,
    190
operator, 17, 25, 32, 33
Exponential distribution, 35, 37, 38,
    41, 46, 63, 67, 68, 99, 105, 113,

114, 132, 134, 145, 157, 169, 190,
    201, 203, 207, 208, 226, 277, 293
Exponential service time distribution,
    106, 108, 148

Fading, 4, 5, 23, 308, 309, 311, 312,
    314-318
Fault-tree, 291
Finite buffer, 98, 115, 116, 171, 214,
    215
Finite capacity queue, 115
First-come-first-served (FCFS), 97,
    104, 105, 108, 113, 114, 117, 120,
    132, 134-136, 142, 145, 148, 150
First-in-first-out (FIFO), 97, 99
FITS, 263, 274
Fixed routing, 213
Flow control, 98, 129, 225, 226, 228,
    230, 233
Flow of probability, 48, 132
Frequency plots, 60, 73
Full-period generator, 54

Gamma distribution, 35
Gate, 260, 305, 306
Gaussian distribution, 34
Generating function, 17-19, 21, 23, 27,
    28, 33, 118, 156, 157, 190, 191
Generating random numbers, 51
Generator
    additive congruential, 55
    full-period, 54
    linear congruential, 52
    linear recurrence modulo 2, 56
    maximal-period, 54
    mixed congruential, 52
    multiplicative congruential, 52, 54,
        56
    period, 53, 55
    prime-modulus multiplicative
        congruential, 54

pseudo random-number, 50-53, 59, 60, 62, 73, 75, 79, 82
Geometric distribution, 23, 110, 191
GI/GI/r queue, 100, 102, 103, 105, 124
Go-back-N protocol, 220, 221

H.261, 242
Half-duplex transmission, 220
Handover, 4, 315-317
Hazard rate, 263, 267
Heavy traffic, 124, 200, 227, 231
Heterogeneous media, 235
High definition television (HDTV), 235

Independence, 14, 15, 25-27, 62, 66, 158, 188, 217
Interarrival time, 96, 105, 116, 123, 124, 203
Inverse transform method, 63

Jackson networks, 132, 133
Joint probability distribution, 25, 132, 140

Kendall's notation, 99
Kingman's heavy traffic approximation, 124
Kleinrock's network delay bound, 213

L'Hopital's theorem, 119
Laplace transform, 33, 118, 120, 135
Last-come-first-served (LCFS), 134
Last-come-first-served(LCFS), 97, 108, 117, 134-136, 142, 146
Last-in-first-out (LIFO), 97
Life testing, 279
Lifetime, 30, 260, 264, 270, 271, 284, 285, 294
    PDF, 270, 271
Light traffic, 226

Linear congruential generator, 52
Linear recurrence generator modulo 2, 56
Link delay, 202, 213, 222
Little's formula, 104, 110, 119, 131, 133, 134, 136, 138, 143-146, 227, 249
Local area network (LAN), 246
Logic gate, 260, 305, 306
Lognormal distribution, 314
Lost traffic, 101, 193

M/D/1 queue, 120, 121, 237
M/D/1/K queue, 237
M/G/1 queue, 73, 247, 248
M/GI/1 queue, 116, 117, 119-121, 123, 124
M/M/1 queue, 111, 112, 114, 119, 135, 201, 202, 213, 214, 226
M/M/r queue, 108, 110, 113, 114, 116, 119, 124, 132, 133, 135
Maintenance, 260, 261, 268, 284, 292, 293, 295, 296
    procedure, 260, 296
    spares, 284
Majority decision redundancy, 287, 295-297, 308
Manufacture, 1, 4, 63, 260, 273, 274, 279, 281, 282, 302
Markov
    chain, 116, 117, 132, 159-161, 164, 165, 171, 244, 245, 263
    model, 159, 161, 183, 244-246, 253, 288, 308
    process, 45, 49, 116, 121, 132, 237, 240, 244, 246, 294, 295
    queue, 108
    two-state process, 240
Maximal-period generator, 54
Mean time between failures (MTBF), 263

Mean time to failure (MTTF), 263, 267, 268, 270, 274, 283, 284, 286, 287, 293, 294
Mean time to repair (MTTR), 263, 293, 294, 296
Mean-value analysis, 145, 146, 151
Memoryless property, 38, 108
Message delay, 202, 208, 211-213
Message switching, 200, 201
Midsquares method, 51
Mixed congruential generator, 52
Mobile radio, 4, 5, 95, 308, 314, 319
Modelling traffic sources, 154, 240
Moments, 17, 18, 33, 119
MPEG2, 242
Multi-component systems, 283
Multimedia services, 235
Multiplicative congruential generator, 52, 54, 56
    prime-modulus, 54
Multiprocessor systems, 149
Mutually exclusive event, 9, 10, 13, 14, 24, 29
Mutually exhaustive event, 13

Negative acknowledgement, 221
Negative-exponential distribution, 35, 37, 38, 41, 46, 51, 63, 64, 67, 68, 96, 99, 105, 106, 113, 114, 132, 134, 145, 155, 157, 161, 169, 170, 189, 190, 201, 203, 206-208, 226, 237, 277, 293
Network
    broadband, 201
    closed, 128, 136, 138, 144, 147
    delay, 208-213
    open, 127, 129, 131, 136, 138
Node, 92, 126, 127, 129-146, 148-150, 200, 203, 226, 232, 234
Normal distribution, 34, 35, 65, 75-77, 241, 244, 270, 271, 280, 303, 304, 306, 308, 314, 316

Normalisation constant, 140-146
Occupancy, 15, 102, 103, 108-110, 117, 119, 124, 133, 135, 137-139, 144, 153, 154, 158, 161, 162, 184, 185
Offered load, 100, 105, 108, 109, 121
Offered traffic, 168, 210, 211, 214-216, 236, 237, 247, 248
Open network, 127, 129, 131, 136, 138
    separable, 132
Optimum packet length, 217, 235

Packet length, 200, 217-219, 222, 226, 235, 249
    optimum, 217, 235
Packet switching, 3, 200, 201, 226, 235, 257, 258
Packet throughput, 218
Pause, 240, 251
Payload, 217, 235, 247, 250, 251
Performance
    assessment, 93
    engineering, 1-3, 5, 6, 35, 93, 259, 264, 302, 307
    measures, 50, 93, 94, 100, 103, 126, 142-145, 148, 150
Poisson
    arrival process, 105, 117, 121, 203
    distribution, 22, 39, 41, 51, 66-68, 105, 118, 120, 121, 157, 161, 170, 217, 226, 237, 284
    process, 38, 39, 41, 42, 44, 46, 67, 68, 95, 96, 102, 105, 106, 108, 117, 132, 189, 203, 240, 244, 245, 250, 255, 263, 270, 282
    queue, 203
Pollaczek-Khinchine formula, 117
Pooled point processes, 203
Population, 44, 45, 60, 62, 75, 95, 99, 265, 267, 280
Power supply, 260

Prime-modulus multiplicative
  congruential generator, 54
Probabilistic design, 306
Probability
  distribution, 16, 19, 34, 51, 63, 66,
    73, 101, 117, 144
  flow, 48, 122, 123, 295
  generating function (PGF), 17-20,
    27-30, 33, 39-41, 118, 120
  measure, 10, 11, 15, 16
  state, 48, 101, 122, 160, 167, 168,
    171, 237, 253, 308
Programming simulations, 69
Propagation, 219, 220, 222, 234, 248,
  314
  delay, 219, 220, 222, 234, 248
Protocol
  automatic repeat request (ARQ),
    218
  go-back-N, 220, 221
  select-repeat, 217, 223, 224
  stop-and-wait, 219, 220, 222, 223
Pseudo random-number generator, 50-
  53, 59, 60, 62, 73, 75, 79, 82

Queue
  discipline, 94, 97-99, 103-105, 108,
    113, 117, 120, 126, 132, 134, 146
  Poisson, 203
Queuing
  models, 96, 99, 101
  systems, 92, 93, 97, 98, 166
  theory, 6, 199, 213, 257

Radio propagation, 219, 220, 222, 234,
  248, 314
Random experiment, 8
Random order of service (ROS), 97,
  108, 117
Random point process, 38, 39, 44
Random variable

compound, 28
continuous, 30-33, 35
discrete, 15-17, 24-26, 30-32
independent, 25-28, 39, 40, 42, 65
Random vector, 24, 132, 140
  discrete, 24
Random-number generator, 50-53, 59,
  60, 62, 73, 75, 79, 82
Random-split routing, 213
Real-time services, 5, 201, 235
Redundancy, 261, 284-288, 291, 294,
  295, 299
  majority decision, 287, 295-297, 308
Redundant system, 294-296
Regenerative simulation, 76
Reliability
  block diagram, 285, 291
  component, 274
  estimation, 264
  function, 265, 267, 269, 287
  software, 298
  system, 93, 274, 286, 288, 294, 296
Repairable system, 291
Repeated transmission, 308
Repeater, 259, 307, 308
Response time, 63, 94, 97, 102, 103,
  112, 113, 133-136, 138-140, 143,
  145, 146, 148, 150, 152
Retransmission, 218, 221
Round-robin scheduling, 98
Round-trip delay, 219, 222
Routing
  fixed, 213
  matrix, 127, 132
  random-split, 213
  short-term adaptation, 213
  static, 213

Sample
  point, 9-11, 15, 31
  space, 9, 11, 15, 24, 25, 27, 29, 31

Sampling, 60, 245, 246, 255, 302
Saturation point, 136, 139, 140
Seed, 51-54, 61, 64, 67, 68, 75, 79, 80, 82, 83, 85, 86
Select-repeat protocol, 217, 223
Server occupancy, 102, 108-110, 119, 133, 137
Service
    facility, 92, 94, 96-98, 104, 126
    system, 92, 95, 99, 121, 126
    time, 35, 36, 78, 94, 97, 98, 102-104, 106-108, 112, 113, 115-121, 123, 127, 131, 132, 135, 136, 139, 145, 148, 150, 159, 201-203, 247, 248
Service-in-random-order (SIRO), 97
Shortest-service-time-first (SSTF), 97
Short-term adaptation routing, 213
Simulation
    event-based, 70, 72, 77
    programme, 51, 69, 70, 73, 76, 77, 89
    programming, 69
    regenerative, 76
    time-based, 70, 71
Sliding window, 225, 226, 228
Software
    fault, 268, 298-300
    reliability, 298
Sojourn time, 102, 104, 115, 120, 133, 144, 145
Source model
    data, 240
    speech, 240
    video, 242, 258
Space switch, 177, 178, 182, 236
Spares, 98, 128, 284, 285, 293
Speech, 3, 201, 207, 235, 240, 241, 250-252, 255-257
Speech sources, 240
Stability, 100, 122, 131, 225, 239
Standby, 288, 290, 291

Start-up conditions, 74, 76, 77
State probability, 48, 101, 122, 160, 167, 168, 171, 237, 253, 308
Static routing, 213
Statistical equilibrium, 47-49, 101-104, 106, 108, 109, 113, 117, 131, 132, 140, 145, 158, 160, 167, 176, 197, 198, 203
Statistical independence, 14, 15, 25-27, 62, 66, 158, 188, 217
Steady-state distribution, 47
Stochastic modelling, 38
Stop-and-wait protocol, 219, 220, 222, 223
Store-and-forward, 98, 200, 226
Student-t distribution, 75, 76
Subscriber's private meter (SPM), 77-79, 81-83, 86, 89
Summary statistics, 73, 101
Switch
    space, 236
Switching
    circuit, 200, 201, 236
    message, 200, 201
    packet, 3, 200, 201, 226, 235, 257, 258
System reliability, 93, 274, 286, 288, 294, 296

Talkspurt, 240, 241, 250
Telephony, 15, 162, 235
Throughput, 100, 129, 136, 137, 139, 142, 145, 146, 148, 149, 215-224, 226-230, 232
    packet, 218
Time assignment speech interpolation (TASI), 240, 257
Time-based simulation, 70, 71
Time-out, 219, 221
Token ring, 246, 247, 249
Token rotation time, 247, 248

Tolerance, 1, 4, 63, 260, 268, 302, 303, 305-307
  component, 303
  timing, 305, 307
Total probability theorem, 13, 24, 29, 47, 114, 252, 254, 255, 310, 311, 313, 317
Traffic
  data sources, 240
  equation, 129, 130, 132, 136, 137, 140, 143
  flow, 159, 195, 225, 234
  heavy, 124, 200, 227, 231
  intensity, 100, 102, 133, 135, 160, 161, 172, 174
  light, 226
  modelling, 154, 240
  offered, 168, 210, 211, 214-216, 236, 237, 247, 248
  sources, 154, 240
  speech sources, 240
  video sources, 242, 258
Transition
  coefficients, 167, 241, 244, 251, 252, 296, 313, 316
  diagram, 45
  matrix, 294, 298
  probability, 288, 295
  rate, 48, 49
Transmission systems, 4, 307, 308

Two-state Markov process, 240

Uniform distribution, 34, 51, 62-64, 154, 155, 157
Unreliability function, 275
Upstream blocking, 214, 215
Utilisation factor, 101

Value
  component, 63, 303
Variable bit-rate (VBR), 240, 258
Verification, 73, 132
Video, 3, 5, 201, 235, 242-246, 257, 258
  source model, 242
Video sources, 242, 258
Virtual channel, 227, 233
Virtual circuit, 230
Visit count, 130, 131

Waiting time
  density function, 206
  distribution, 102, 103, 113, 114, 120, 133, 206
Wear out, 268
Weibull distribution, 271-273, 276-278, 300
Whitt's approximation, 124
Window size, 225, 226, 229-232, 234
Worst case design, 305

Printed in the USA
CPSIA information can be obtained
at www.ICGtesting.com
JSHW011518221024
72172JS00008B/59

$$P(A) = \sum_{i=1}^{n} P(A\&B_i) = \sum_{i=1}^{n} P(A|B_i) P(B_i)$$

(2.4)

The theorem is established by considering A as the mutually exclusive union of the events A & B_i and appealing to axiom (iv).

2.2.3   *Statistical independence*

Events A and B are statistically independent if

$$P(A|B) = P(A)$$

i.e. knowing that B has occurred on a particular trial does not affect the probability that A also occurred. Since P(A|B) = P(A&B)/P(B), we deduce that if A and B are independent events.

$$P(A\&B) = P(A) P(B)$$

(2.5)

This special multiplication rule is sometimes used to define independence.

Statistical independence is an important concept, because if it can be proven, or more usually assumed, to hold between various events, then the above multiplication rule considerably simplifies the calculation of probabilities.

> *Example: binary data is transmitted in blocks of 32 bits down a noisy channel which causes 3 % of bits transmitted to be received with the wrong value. What is the probability that a single block contains one or more transmission errors?*

Let A_i be the event 'ith bit of block received correctly' $(1 \leq i \leq 32)$

Then $P(A_i) = 1 - 0.03 = 0.97$.

Let A be the event 'whole block received correctly'.

Clearly, $A = A_1 \& A_2 \& \ldots \& A_{32}$.

If we assume that individual bit transmission errors are mutually independent, then the probability that the whole block is correctly received is